Careers in Engineering and Technology

Careers in Engineering and Technology

W. EDWARD RED
Brigham Young University

Brooks/Cole Engineering Division
Monterey, California

Brooks/Cole Engineering Division
A Division of Wadsworth, Inc.

Printed in the United States of America

10 9 8 7 6 5 4 3 2 1

Library of Congress Cataloging in Publication Data

Red, W. Edward.
    Careers in engineering and technology.

    Bibliography: p.
    Includes index.
    1. Engineering—Vocational guidance.   2. Technology—
Vocational guidance.   I. Title.
TA157.R37   1984   602.3   83–24059

ISBN 0-534-03141-2

Sponsoring Editor: Ray Kingman
Manuscript Editor: Janet Greenblatt
Production: Ex Libris □ Julie Kranhold
Interior Design: Sara B. Hunsaker
Cover Design: John Edeen
Cover Photo: Courtesy of PDA Engineering, Santa Ana, California
Typesetting: Interactive Composition Corporation
Printing and Binding: R. R. Donnelley & Sons
Production Services Manager: Bill Murdock

Preface

Careers in Engineering and Technology will help students decide if their interests and talents are compatible with an engineering career and will help them select a branch of engineering or, for technology students, a technology discipline. Although this book focuses on engineering careers, it recognizes that strong programs in engineering technology have emerged in the last decade to provide technical alternatives for those students less conceptually inclined but more hardware and applications oriented. The seven chapters of this book are structured into three parts:

- **Part One:** Engineering and Technology
- **Part Two:** Preparing for a Profession
- **Part Three:** Fundamental Engineering Skills

Part One introduces students to the engineering environment, delineating the career flexibility offered in the numerous engineering branches and engineering functions. The activities involved in the engineering functions are discussed as they relate to the engineering design process and as they involve the different members of the technological team. Several case studies in engineering design are incorporated to demonstrate the type of design situations that engineers and technologists may face.

The material on engineering and technology education examined in Chapter 3 of Part Two provides practical guidelines for maneuvering through engineering and technology curricula. This chapter also covers the transitional period from graduating student to experienced professional, including material on résumés, job interviews, plant trips, continuing education, and professional societies.

Chapter 4 of Part Two covers the professional responsibilities associated not only with engineering but with all professions. It includes two case studies in ethics to stimulate interesting class discussions in the areas of personal responsibility and professional integrity. Chapter 4 concludes with the *Code of Ethics* of the National Society of Professional Engineers.

Part Three introduces fundamental, yet important, skills required of students in engineering and technology. This material may not be necessary for students already skilled in the use of hand-held calculators (Chapter 5) or those who are well versed in the use of different dimension and unit systems (Chapter 6). Nevertheless, it is recommended that students demonstrate their proficiency by working the problems at the end of the chapter before skipping this material.

Chapter 7, the concluding chapter of the book, introduces students to one of their most valuable resources for continuing education, the technical library. This chapter also introduces the topic of patents and concludes with an interesting case study that illustrates patent complexities. I am indebted to technical librarians Barbara Hedges, Paula Higgins, and Julia Rholes for their significant contributions to this chapter.

I am indebted to the numerous engineering industries, publishing companies, and individuals who have satisfied my requests for tables, diagrams, and photographs or have granted permission to use excerpts from books and articles. I also acknowledge the efforts of Zina Niemeyer for assembling the final manuscript.

W. Edward Red

Contents

PART ONE

Engineering and Technology 1

1 Engineering in the Twentieth Century 2

Engineering Defined 3
Branches of Engineering 4
Aerospace Engineering 6
Agricultural Engineering 9
Chemical Engineering 10
Civil Engineering 13
Computer Engineering 16
Electrical Engineering 20
Industrial Engineering 23

Materials and Metallurgical
Engineering 25
Mechanical Engineering 27
Mining Engineering 31
Nuclear Engineering 32
Petroleum Engineering 33
Engineering Technology 35
Problems 37

2 Design and the Engineer's Function 38

Engineering Design 39
The Design Process 44
The Phases of Design 45
The Technological Team 48
Engineering Functions 50
Research: Knowledge, Understanding,
Application 52
Development: Methods, Devices,
Performance 55
Design: Selection, Specification 60

Production and Construction: Processes,
Assembly, Structures, Systems 65
Operations and Maintenance:
Performance, Service, Repair 70
Sales: Marketing, Applications,
Service 73
Management: People, Projects,
Products 75
Looking Back 79
Problems 80

PART TWO

Preparing for a Profession   83

3 Pursuing a Career in Engineering or Technology   84

Is Engineering or Technology for Me? 85
The Education 86
Preparing for a Profession 101

Entering the Profession 104
The Professional Years 112
Problems 115

4 The Profession   116

Professionalism: Characteristics and Responsibilities 117

Cases in Ethical Studies 124
Problems 136

PART THREE

Fundamental Engineering Skills   139

5 The Hand-Held Calculator   140

Introduction 141
Questions about Calculators 141
Basic Features of a Calculator 144
Essential Calculator Functions 148
Nonessential but Useful Calculator Functions 152

Algebraic Operating System (AOS) 153
Reversed Polish Notation (RPN) 156
Some Common Errors 158
Practical Examples 159
Problems 161

6 Dimensions and Units 162

Historical Background   163
Dimensions   166
Unit Systems   174
International System of Units (SI)   174

Gravitational Unit Systems   183
Conversion of Units   183
Problems   188

7 The Technical Library 192

The Technical Library   193
Reference Works   197
Standards and Specifications   203
Government Documents   204
Technical Reports   206

Two Reference Searches   207
Patents   211
The Drama of Patents   214
Problems   222

Appendixes 224

A: Conversion Factors and Other Constants   224

B: Library Reference Materials   239

Selected Bibliography 246
Answers to Problems—Chapters 5 and 6 249
Index 251

Careers in Engineering and Technology

PART ONE

Engineering and Technology

CHAPTER 1
ENGINEERING IN THE TWENTIETH CENTURY

CHAPTER 2
DESIGN AND THE ENGINEER'S FUNCTION

Engineering in the Twentieth Century

CONTENTS

Engineering Defined
Branches of Engineering
Aerospace Engineering
Agricultural Engineering
Chemical Engineering
Civil Engineering
Computer Engineering
Electrical Engineering
Industrial Engineering
Materials and Metallurgical Engineering
Mechanical Engineering
Mining Engineering
Nuclear Engineering
Petroleum Engineering
Engineering Technology

Engineering activities now reach into every area of society, wherever problems need solving. In less than a century, the number of established engineering branches has multiplied fivefold. The future promises still more expansion as the unique talents of engineers continue to be recognized and needed, and as technological innovation continues to create new areas of technical expertise.

ENGINEERING DEFINED

A pamphlet distributed by the Engineers' Council for Professional Development (ECPD), now the Accreditation Board for Engineering and Technology, Inc. (ABET), gives a classic definition of engineering:

> The profession in which a knowledge of the mathematical and natural sciences gained by study, experience and practice is applied with judgement to develop ways to utilize, economically *and with concern for the environment and society,* the materials and forces of nature for the benefit of mankind. [Emphasis supplied.]

The italicized phrase distinguishes this definition from the one adopted by the ECPD in 1963. The insertion of "with concern for the environment and society" obviously hints at the sensitivities and reservations that surround engineering and technology today. We live in an age when technological achievements can quickly influence the living standards of individuals, nations, and societies, yet these achievements sometimes exact a high price from our natural resources, our environment, and our cultural heritage.

Some obvious examples of the changes wrought by technology can be found in transportation and communications. The world has become a smaller place now that affordable automobiles, rapid transit systems, jet airliners, and satellites have made travel and communication between continents a matter of hours—or even seconds. This easy interchange has contributed to an unprecedented mingling of cultures, and in many countries the multinational society is becoming a reality.

At the same time, demands on world resources have made nations dangerously competitive while straining nature's ability to modify environmental side effects that accompany the use of these resources. It has become increasingly apparent that the impact of any new technology must be carefully

3

To work in harmony with nature is perhaps the greatest challenge to engineering to-day: to preserve and protect the quality of life while developing our resources for the benefit of humankind. This beautiful lake, the product of engineering knowledge and skills, required the construction of the dam shown in the following picture. (Courtesy of U.S. Army Corps of Engineers)

evaluated *before* it is applied. Today's engineers must not only continue their familiar role of raising living standards by meeting human needs but are now asked to solve problems that their predecessors unknowingly caused.

BRANCHES OF ENGINEERING

Prior to the Industrial Revolution, engineers were primarily "builders" whose activities paralleled those of modern civil engineers. With the advent of the Industrial Revolution came the need for more engineers having different interests and abilities. Spurred on by the rapidly expanding base of scientific knowledge, industrial technology began to produce goods, devices, systems, products, and structures. These proved to be self-stimulating; technology began to feed on itself as new knowledge and industrial applications served to identify new needs that in turn required still further technological advance-ment. This increasing role of technology led to the need for new types of engineers.

Only time will tell whether this solution to a water-resource problem is in the best interests of humankind. (Courtesy of U.S. Army Corps of Engineers)

By the turn of the twentieth century there were six branches or fields of engineering (there is a tendency to refer to the "branches" of engineering as "fields," even though the word *branch* is more proper): agricultural, chemical, civil, electrical, mechanical, and the branch broadly comprising mining, metallurgical, and petroleum engineers. Today degrees are offered in some twenty-five to thirty separate engineering curricula, most of which we can classify into various branches of engineering. The more common ones are:

Aerospace	Electrical	Materials/Metallurgy
Agricultural	Energy/Power	Mechanical
Architectural	Engineering Sciences	Mining
Biomedical	Environmental	Nuclear
Chemical	Geological	Petroleum
Civil	Industrial	Systems
Computer	Marine/Naval	

From this group, electrical engineering will graduate about 27 percent of the total engineering graduates in any one year; civil engineering, 20 percent;

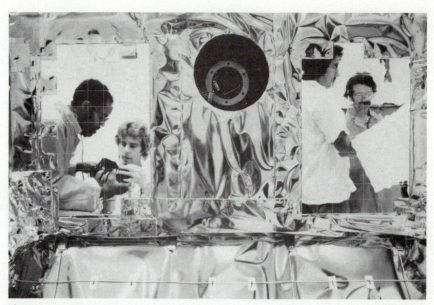

Engineers, technologists, and technicians, mirrored in the solar panels of a satellite, are working together as a technological team to ready the satellite for shipping and launch. (Courtesy of TRW)

mechanical engineering, 18 percent; chemical engineering, 9 percent; and industrial engineering, 6 percent. These branches alone account for approximately 80 percent of the engineering students being graduated annually.

New branches of engineering usually originate in one of two ways: (1) as derivatives from a specialty area in an established branch of engineering, or (2) from an area of society finding itself newly dependent on technology. Some of the most obvious examples of the first are the now-separate branches of mining, metallurgy, and petroleum engineering.

One example of a new branch of engineering, established primarily by introduction of technology into the life sciences of medicine and biology, is **biomedical engineering.** The increasing need for and dependence on instrumentation and mechanical devices to monitor and control life functions gradually created the need for engineering specialists, which eventually led to the development of this new branch of engineering.

The following sections briefly examine some of the branches of engineering, including specialty areas common to each branch.

AEROSPACE ENGINEERING

Aerospace engineering is concerned with engineering applications in the areas of **aeronautics** (the science of air flight) and **astronautics** (the science of space flight). Aerospace engineering thus deals with such varied vehicles of flight as balloons, sailplanes, propeller- and jet-powered aircraft, missiles, rockets, and satellites, as well as advanced interplanetary concepts like ion propulsion rockets and solar wind vehicles. It is an exciting and challenging field in which engineers are applying and advancing the state of the art in the sciences and in engineering. It has changed the face of the globe, as was predicted by Charles Dollfus, the noted French historian, after the successful flight of Wilbur and Orville Wright in 1903:

> It is therefore uncontestably the Wright brothers alone who re-solved in its entirety the problem of human mechanical flight. . . . Men of genius—erudite, exact in their reason, hard workers, out-standing experimenters and unselfish. . . . They changed the face of the globe. (Gibbs-Smith, 1978, p. 15.)

Some of the major efforts of aerospace engineers occur in the specialty areas of *aerodynamics, structural design, instrumentation, propulsion systems, materials, controls, life systems, reliability testing, navigation, production methods, cryogenics* (the study of material properties at low temperatures), *thermodynamics,* and *avionics* (communications and flight control). Obviously, many of these fall into the specialty areas of other fields of engineering, allowing engineers of all kinds to participate in aerospace engineering. What distinguishes aerospace education from other fields of engineering is its extension over many areas.

Nowhere else have the relationships among the members of the technological team (the engineer, scientist, technologist, technician, and craftsperson) been so fully developed. Meeting the continuing challenges of conquering space and providing better defense systems as well as faster, more reliable means of transportation and communication have required large expenditures of money and personnel in research and development (R&D). Because of this, nearly 25 percent of aerospace employees are engineers, scientists, and technicians, a much higher percentage than is used in other technical industries.

Few people comprehend the number of useful discoveries, products, and by-products that have resulted directly or indirectly from aerospace technology, some of which are revolutionizing our world today. Among them are solid-state integrated circuitry, minicomputers and microprocessors, fuel cells, solar cells, laser guidance and control systems, life-support systems, satellite communication systems, and a host of improved materials now used in various industries. In fact, the pace of discovery is so fast that materials and products often become obsolete before production is completed.

A review of the history of the aerospace industry reveals periods in which engineering employment varied significantly. In general, this resulted from huge national projects, such as Apollo, Minuteman Missile, and SST, which employed large numbers of technical people. Many engineers became temporarily unemployed when government financial support was inevitably curtailed upon the completion of the major project tasks. Nevertheless, these engineers eventually found work in other industries or returned to the aerospace industry. It now appears that the aerospace industry as a whole is more

The Wrights' first airplane engine was a 12-horsepower, four-cylinder engine that weighed 179 pounds. (Courtesy of the Science Museum, London)

Wilbur Wright's first attempt at flight on December 14, 1903, ended when he overcorrected with the elevator and the "Flyer" plowed into the sand. (Courtesy of the Science Museum, London Crown Copyright)

stable because of increased diversification. At the same time, the government is less likely to fund projects at the previous levels. Consider, for example, the space shuttle project authorized by President Nixon in January 1972. Funded at an annual rate averaging about $4 billion to $5 billion and spread over a decade, the shuttle program, by virtue of the reusable orbiter, promises savings of almost 30 percent over that required by expendable launch vehicles to operate in space. This program also is providing many smaller industries, educational institutions, and other groups their first access to space.

In this artist's conception you can see bricks of pure silica. These bricks protect NASA's space shuttle from the intense heat that turns much of the underbelly white-hot during re-entry into the earth's atmosphere. (Courtesy of Lockheed Corporation)

AGRICULTURAL ENGINEERING

At the turn of the century, a large percentage of the working population was engaged in agriculture—quite a contrast to today's 5 percent. This startling change is attributed to the integration of technology and engineering into agriculture, enabling modern farmers to feed approximately ten times as many families as their ancestors did a hundred years ago. Recognized as the world leader in agricultural technology, the United States exports grains, vegetables, wood, and by-products as well as agricultural machinery and systems technology to many foreign countries.

There are five major specialty areas within agricultural engineering: *soils and water, structures and environment, electric power and processing, food engineering,* and *power and machinery.*

SOILS AND WATER   Agricultural engineers who work with soils and water are concerned with erosion control, surface and subsurface water drainage, large- and small-area irrigation by canals and sprinklers, structures for water retainment, land-use methods, water and land conservation, and sewage disposal.

STRUCTURES AND ENVIRONMENT   This specialty will attract engineers interested in the analysis, design, and development of a variety of agricultural structures, such as housing for livestock and poultry; storage structures for products such as rice, grain, corn, and beans; and, in smaller numbers, greenhouses to grow plants and vegetables. Familiarity with materials and the ability to determine structural loads and strengths are necessary. Proper design also involves the maintenance of a proper environment to prolong the life of stored products; thus, water, light, noise, air quality, temperature, sanitation, and waste have to be monitored and controlled. Solar technology is rapidly becoming prominent in this area.

ELECTRIC POWER AND PROCESSING   This area deals with the application of electric energy to the many areas of production and processing. This may include electrical usage in milk-processing equipment, grain and feed processing, farm lighting, systems control, and farm materials handling. In addition, radiation treatment is now being used by agricultural engineers for food and container sterilization and for food preservation.

FOOD ENGINEERING   Engineers in this field develop improved food-handling procedures and equipment that reduce handling damage and waste, improve sanitation, are less costly, and conserve raw materials and energy.

POWER AND MACHINERY   More agricultural engineers are employed in this area than any other, particularly in the design and development of agricultural equipment such as tractors, harvesters, loaders, fertilizer spreaders, animal feed distribution systems, and forage boxes. They may develop better hydraulic systems or improve the design of handling and processing machinery. They may be interested in the interaction of the equipment with the soil being farmed—meaning they must understand soil types, soil dynamics, and traction efficiency. Equipment will have to be tested and proved reliable while providing safety and comfort to those using it. Many engineers will work in **forest engineering** by developing equipment to cut, utilize, replant, and manage our forests, one of our most valuable resources.

Agricultural engineers do not necessarily live in rural areas; in fact, most will work in cities, where the majority of our farm equipment manufacturers are located. Their educational background will be similar to that of other kinds of engineers, including a basic knowledge of the maths and sciences and a study of fundamental engineering subjects. As the United States continues to lead the way in agricultural technology, the future for agricultural engineers appears bright.

CHEMICAL ENGINEERING

Chemical engineering is the application of chemical science to industrial processes that change the composition or properties of matter for useful purposes. Some of the common applications are found in the manufacture of chemicals (including drugs, cements, paints, lubricants, pesticides, fertilizers); in oil refining, combustion, extraction of metals from ores; and in the production of ceramics, brick, and glass. In addition, chemical engineering is involved with food processing, petrochemical products, nuclear energy,

In this machine, food engineers test the effects of a new vacuum-drying technique on vegetables. (Courtesy of NASA)

coal gasification, coal liquefaction, and control of pollution involving air, water, and waste solids.

Chemical engineers can apply their skills in *food engineering, process dynamics and control, environmental control, electrochemical engineering, energy production, polymer science technology, unit operations,* and *plant design and economics.*

FOOD ENGINEERING Chemical engineers working with foods will develop better sterilization methods, improve food additives, and utilize protein extracts from plants.

PROCESS DYNAMICS AND CONTROL This is an important specialty for engineers interested in developing and controlling chemical processes and materials, many of which are highly sensitive to environmental infringement or fluctuation. This specialty becomes even more important when one considers the size and complexity of our industrial plants.

ENVIRONMENTAL CONTROL Because production of energy and of chemical products often leads to undesirable and environmentally harmful

Chemistry is naturally an important subject for chemical engineers. Thus, chemical engineering students should be familiar with the equipment of a chemistry laboratory and with fundamental chemical reactions. Nevertheless, the everyday work of the chemical engineer is quite different from that of the chemical scientist because of the innate differences between engineers and scientists. (Courtesy of Monsanto)

pollutants, the need for chemical engineers in pollution control has increased rapidly. Working in this area will require a knowledge of microbial and enzymatic activities, neutralization, flotation, design, economics, process treatments, and air, soil, and water ecology. Engineers will develop techniques and equipment that control and detect pollution. Some examples are scrubbers, cyclones, electrostatic precipitators, activated sludge, trickle filters, direct burial techniques, and holding ponds.

ELECTROCHEMICAL ENGINEERING This specialty uses electricity to cause or prevent chemical reactions. Of the total power generated in the United States, almost 10 percent is estimated to be consumed by the electrochemical-processing industry. The major work of electrochemical engineers is to create electrolytic deposition of materials, to prevent corrosion of metals, and to generate electricity from chemical reactions, as in common batteries.

ENERGY PRODUCTION The worldwide shortage of energy is expected to increase the need for chemical engineers to work with the chemical processes involved in producing energy from nuclear reactors, solar devices, coal gasification, coal liquefaction, and, of course, petroleum. Chemical engineers must be familiar with thermodynamic principles, material reaction to high temperatures, and radiation environments.

POLYMER TECHNOLOGY   Approximately 30 percent of all chemists and chemical engineers are employed in this area. **Polymerization**—joining two or more like molecules to form a more complex molecule having different physical properties—is responsible for many new synthetic materials, such as polystyrene, nylon, silicone rubber, phenolic laminate and composites, thermoplastics, polypropylene, polyethylene, and polyvinyl chloride. These materials have revolutionized our textiles, containers, fasteners, and structures. In many industries, synthetics have replaced expensive and sometimes functionally limited traditional materials such as wood, metals, rubber, and glass.

UNIT OPERATIONS   This field of chemical engineering is concerned with fluid transportation through ducts or pipelines, solid material transportation by conveyors or fluid media, heat transfer from one substance to another, absorption of gases, and distillation recovery of products. Chemical reactions such as chlorination, oxidation, hydrogenation, reduction, nitration, sulfonation, and pyrolysis will often be used in these operations to produce new materials.

PLANT DESIGN   Chemical engineers in this area will apply their knowledge of systems and controls, along with basic economic principles, to the layout of industrial plants. Since they must interact with managers, construction personnel, plant operators, and engineers from other fields, a well-rounded background requiring years of experience will be necessary.

Chemical engineers can expect to work in a highly competitive environment in which new ways of applying chemical synthesis are constantly being developed. This is in sharp contrast to the early part of the twentieth century, when the rate of obsolescence of chemical products was low. However, the chemical products of that time—acids, alkalies, chlorine, salt components, and fertilizers—are still very much in demand.

CIVIL ENGINEERING

Civil engineering, by far the oldest engineering branch, is primarily concerned with structures such as high-rise buildings, recreational facilities, housing, industrial plants, dams, nuclear facilities, boats, ships, railroads, rapid transit systems, tunnels, satellites, rockets, harbors, offshore oil and gas facilities, pipelines, canals, bridges, and highway systems. The failure of any of these structures can severely disrupt the well-being of society. Thus, civil engineers must assume a larger measure of responsibility for their actions than most other professionals.

Like other kinds of engineers, civil engineers usually work in specialty areas that have arisen from the new technology needed to support the increased versatility and complexity of our modern structures. The main specialty areas are *aerospace, construction, energy, sanitation and environment, geotechnology, hydraulics and water resources, mechanics, structures,* and *urban planning, highways, and transportation.*

AEROSPACE   Civil engineers in aerospace are often involved in the construction of airplanes, airports, rockets, missiles, satellites, and launch facilities. This may require the investigation of the effects of weightlessness on structural configurations or the selection of new materials that can survive an environment in which severe changes in temperature or loading may be caused by vibration. They will work closely with other engineers, scientists, technologists, and technicians as they meet the unique challenges found in aerospace.

CONSTRUCTION   This field offers engineers many opportunities to work outdoors and with different people; for example, construction personnel, architects, people in crafts trades, politicians, and representatives of city government. Engineers are usually involved in the preliminary planning, cost estimating, surveying, excavation, and layout of buildings, dams, plants, and so forth. They may be involved in cost control and contract administration and must have the ability to work well with government inspectors and engineering subcontractors.

ENERGY   The increasing demand for energy has made this area an important one for civil engineers, as new facilities and methods have been required for more efficient resource utilization. Power plant and nuclear facility design, construction, and siting will occupy the efforts of many engineers in this area, as will establishing and interpreting regulations and policies governing these types of facilities. In addition, engineers may design pipelines, solar structures, or wind towers or may work to develop building techniques and materials that will conserve energy.

SANITATION AND ENVIRONMENT   Engineers in this area develop methods, build facilities, and establish policies for the control, reclamation, and disposal of environmentally harmful wastes such as sewage. In some environmentally sensitive urban areas, civil engineers may have to develop storm-water treatment systems and other methods to minimize the discharge of effluents and contaminants into rivers, lakes, and larger bodies of water. In mining areas, environmental engineers will be engaged in reclamation and in the prevention of subsidence and mineral tailings contamination. Of necessity, sanitary and environmental engineers will be well schooled in chemistry.

GEOTECHNOLOGY   Geotechnical engineers are concerned with the support of structures in earthquake regions. This requires a knowledge of foundations, soil properties, and soil behavior in static and seismic environments. In addition, geotechnical engineers will develop and apply underground mining and storage techniques.

One of the largest and most challenging engineering triumphs of recent years is this Ekofisk complex in the middle of the North Sea. Can you imagine the design problems that civil engineers had to solve before drilling could even be started? (Courtesy of Phillips Petroleum Company)

HYDRAULICS AND WATER RESOURCES Irrigation and water management, major concerns of the earliest engineers, are the responsibility of engineers who are charged with managing and developing our most important and increasingly limited resource, water. They will study tidal currents and the flow of water through ditches, conduits, and dams; they will also be concerned with water seepage, erosion, and evaporation. Using this knowledge, many engineers will engage in water management projects requiring irrigation systems and dams.

MECHANICS Civil engineers who specialize in mechanics are interested in understanding the engineering properties of materials and structures in static and dynamic environments. They may investigate soil, rock, and concrete properties; they apply advanced analytic techniques to investigate structural vibrations and the response of structures in dynamic environments such as wind loading. Some civil engineers are involved in investigations of the mechanics of the human body, an area called **biomechanics.**

STRUCTURES Here, engineers are responsible for the design, development, selection, and installation of structural systems and components. Common components such as reinforced concrete, trusses, beams, cables, and laminated timbers will be designed to resist weathering while providing strength, stability, and uniformity.

URBAN PLANNING, HIGHWAYS, AND TRANSPORTATION Civil engineers play an important role in this area. City planners develop plans for the expansion of cities into surrounding rural areas by surveying and mapping these areas, by planning for residential and industrial growth, by laying out streets, by establishing zoning restrictions, and by identifying the public

This is an industrial facility for waste treatment. Industries now spend considerable money and effort in minimizing the waste from their production and manufacturing activities. Here it is cleansed and neutralized before being discharged. (Courtesy of Pratt and Whitney Aircraft Group)

facilities that are needed to support the predicted growth. Highway engineers also support urban growth by planning the system of highways and interchanges that have become an integral part of modern transportation. Engineers interested in all phases of transportation will design and build airports, railroads, subways, and rapid transit systems.

The diversity of civil engineering provides opportunities for civil engineers to work in almost any country with almost any size company. They will have opportunities to work with a variety of people from the public sector and act in a supervisory capacity on many projects. Because many projects are federally or municipally related, many civil engineers are licensed professional engineers.

COMPUTER ENGINEERING

The role that computers play in engineering design has been increasing steadily since the early 1960s. Today, it would be extremely rare to find an industrial organization, large or small, that does not have some modern computational device or system. All kinds of organizations now depend on digital computers to manipulate and store data. In fact, the adoption of computers has increased especially rapidly in nonindustrial sectors of society, particularly in business administration where inventories, financial management, and personnel records require that large amounts of data be readily accessible and easy to manipulate.

Data acquisition requires sophisticated and expensive computer equipment and other instrumentation. This is the control room of the testing center of a large aerospace company. (Courtesy of the Boeing Company)

In technology the trend toward computer utilization is equally impressive. The speed, accuracy, and versatility of the modern computer have made it possible to solve problems once considered intractable and to provide virtually instant access to the solution. The rapidity with which data can be acquired and processed means that systems can be tested, analyzed, and controlled in **real time:** in other words, in their normal operating environment. Most new control systems operate in real time. Automobiles now use microprocessors to control combustion processes; for greater engine efficiency, most aircraft operational systems are controlled by on-board minicomputers; manufacturing and production processes and equipment are controlled by mini- and even microcomputers. The list of computer applications in industrial technology is almost unlimited.

The tremendous growth of this technological tool has led to a new branch of engineering called **computer engineering.** Being relatively new, computer engineering is not offered as a separate curriculum at all universities. But where it is not offered separately, engineering departments, particularly electrical engineering departments, offer it as a specialty study.

Both practicing computer engineers and engineering students are concerned with the **hardware** and the **software** of computers. The term *hardware* describes the physical equipment of a computer, including the electronics components; the mechanical elements used in the complex control systems; and the input/output devices such as card readers, magnetic tapes, printers, plotters, and graphics terminals.

In recent years, engineers have reduced the overall size of computers through developments in microelectronics. Computers are now classified according to size as maxicomputers, minicomputers, and microcomputers. Prices of microcomputers have declined to the point where they are sold as "personal" computers for home use and small-business applications.

In this computer graphics work station we see the refresh cathode-ray tube (CRT), the light pen, the function keyboard, and the alphanumeric keyboard. (Courtesy of Domglas, Inc.)

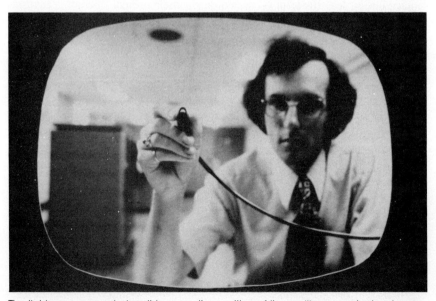

The light pen uses a photocell to sense the position of the oscilloscope electron beam in the cathode-ray tube by closing an electric circuit. The position can then be used to represent a graphical item by interactive manipulation of the input and output data to and from the computer. (Courtesy of Domglas, Inc.)

Although mostly concerned with hardware development, computer engineers must be knowledgeable about the software used to process information. *Software,* as the logic by which computers operate is commonly called, comprises number systems, data formats, programming languages, and operational system logic. Computer engineers will usually comprehend several programming languages like BASIC, FORTRAN, and Pascal. In addition, they will be familiar with the advanced mathematical concepts in subjects like linear algebra, matrix theory, topology, and Boolean algebra. Often they will work with computer scientists and mathematicians as they coordinate new developments in hardware with the advances in software logic.

Computer-Aided Engineering

The newest trend for **computer-aided engineering** (often referred to as CAE) seems to be in the areas of computer-aided design (CAD) and computer-aided manufacturing (CAM); some have even called it a revolution. The rapid decrease in the cost of minicomputers, graphics terminals, plotters, and printers now allows designers to refine their designs *interactively* by interpreting them visually and then modifying them as necessary. The basis of this human/machine communication is the eye. For most individuals, over 70 percent of all information received by the brain is visual. Thus, the ability of designers to synthesize and analyze is obviously enhanced when designs are translated into graphic images on a screen.

Traditionally, design specifications are transmitted to production, construction, and manufacturing personnel to be turned into drawings, blueprints, and specification details. In the past, delays in completing the design specification stage impeded attempts to increase design productivity. Any changes in the design required designers to spend many, many hours—even weeks—revising drawings and associated details. With the adoption of interactive CAD, design productivity has been multiplied by two, ten—or even more. It is now common for integrated circuits, machine elements, piping systems, structures, and so forth, to be designed by CAD methods and then to have the appropriate drawings automatically generated on plotters. In addition, there are computer programs that can take the final description of the design and generate the required manufacturing procedures. For example, some interactive graphics programs are currently being employed to generate the tapes that control the machining motion of numerically controlled (NC) machine tools.

Computer-aided engineering, and particularly interactive graphics, will no doubt become an even more important problem-solving and design tool for engineers in the future. It is hoped that it will make major inroads into the design process by reducing the number of design iterations required to achieve a satisfactory design solution. Engineering students can expect to be introduced to this new design tool in their studies.

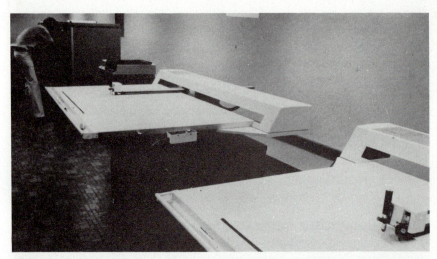

The final output of the system comes from plotters. Once a design has been completed, a permanent-ink, hard-copy drawing is made. (Courtesy of Domglas, Inc.)

The final hard-copy drawings can be used to yield excellent quality reproductions. (Courtesy of Domglas, Inc.)

ELECTRICAL ENGINEERING

Electrical engineering, the largest branch of engineering, employs approximately 250 000 electrical engineers. This is attributable to the widespread utilization of electricity in the areas of transportation, communication, industrial research, development, and production and in the multitude of systems and devices that have become an integral part of modern society. In this expanding environment, electrical engineers have almost unlimited opportunities to apply their skills.

Among the major specialty areas in electrical engineering are *electronics and solid-state circuitry, communication systems, computers and automatic control, instrumentation and measurements, power generation and transmission,* and *industrial applications.*

ELECTRONICS AND SOLID-STATE CIRCUITRY   Electrical engineers have moved electronics into a new era by using solid-state circuitry to develop new devices and components that for all practical purposes have replaced the old space-consuming vacuum-tube technology. Microcircuitry is assisting the revolution occurring in information systems, instrumentation, and controls. Perhaps the most notable development is the microprocessor. Microprocessing technology promises computational and control devices that are orders of magnitude more compact than their predecessors, much cheaper, and more accurate and reliable. Engineers who work in electronics design and development will require a knowledge of both electrical science and materials.

COMMUNICATIONS SYSTEMS   Electrical engineers in this area work to improve communication equipment in radio, telephone, telegraph, television, and defense systems, as well as in other areas. A knowledge of antennas, microwave and sound wave propagation, lasers, electromagnetic principles, and electrical properties of materials may be required.

COMPUTERS AND AUTOMATIC CONTROL   The processing and control of information is the work of modern computers. One of the major technological achievements of the twentieth century, computers provide job opportunities for large numbers of electrical engineers, who apply mechanical and electrical principles to the design and construction of computers of all sizes. Although first developed to support scientific and engineering endeavors, computers are now employed by all industries requiring data processing. For example, they are applied in the automatic control of machines such as milling machines and in autopilots for missiles, ships, and planes. They are used for controlling the many production systems found in industrial plants and in the data processing necessary in banks, businesses, and government.

INSTRUMENTATION AND MEASUREMENTS   These have become increasingly important as we seek to control and measure the various processes around us. Electrical engineers develop and utilize equipment that will detect and monitor vibration, temperature, stress, strain, voltage, current, radiation,

Microprocessors are becoming an integral part of control systems. Electrical engineers have assisted in the tremendous developments in microcircuitry. This microprocessor contains a sixteen-bit internal architecture. (Courtesy Intel Corp. and Regis McKenna, Inc.)

and the many other properties that conveniently allow us to describe the dynamic environment about us.

POWER GENERATION AND TRANSMISSION   The major area in electrical engineering, the production and transmission of power promises future challenges as energy consumption shifts from petroleum to electricity generated by nuclear and coal-fired plants and hydroelectric dams. Some electrical engineers may be engaged in technically innovative energy conversion by solar, fuel cell, or wind generators. Some may improve the power transmission and insulation characteristics of materials, whereas others may be concerned with developing more efficient power generators or with increasing the energy efficiency of appliances and illumination devices.

INDUSTRIAL APPLICATIONS   Many engineers will discover new applications for electrical energy in industry, government, and the home. This will result in new and improved home applications, heating and cooling systems, transportation systems, calculators, computers, typewriters, measurement devices for industry, health care equipment, and so forth. Since discoveries of new applications often lead to industrial expansion, applications engineers often can work with young, developing companies.

Because the application of electrical energy seems almost unlimited, electrical engineering will continue to diversify, providing new and unique opportunities for future electrical engineers.

Electrical engineers and electrical technicians often work with sophisticated instrumentation to test integrated circuits. (Courtesy of TRW)

INDUSTRIAL ENGINEERING

Industrial engineering is the application of principles and techniques to the design, installation, and improvement of systems that integrate people, materials, and equipment to provide efficient production of goods. Industrial engineering requires knowledge and skills in the mathematical, physical, and social sciences, and in engineering analysis and design in order to specify, predict, and evaluate the results to be obtained from these industrial systems.

The need for industrial engineers has arisen because of the growth in industrial technology that subsequently has increased management and production complexities. The improvements in industrial operations made by industrial engineers have been crucial to company success in a highly competitive world marketplace.

The main specialty areas in industrial engineering are *management engineering, manufacturing engineering, plant design, production and quality control, data processing, and systems analysis.*

MANAGEMENT ENGINEERING It is not unusual to find industrial engineers involved in the leading management of industrial systems. Since management is people oriented, engineering managers may involve themselves in project evaluations, development of wage-incentive systems, project planning and control, and employee benefits. They may work with vendors, subcontractors, labor personnel, and other engineers and managers. Since industrial success is related to return on investment, managers must be well versed in financial management. Legal and economic principles will be applied to the consideration of bids, quotations, mortgages, licenses, easements, tenancy agreements, patents, break-even studies, value analysis, and so forth.

MANUFACTURING ENGINEERING An idea becomes a product through research and development and subsequent manufacturing. Industrial engineers in this area become involved with arranging for production. The activities of systematic planning, design, and arrangement of processing methods and equipment will be conducted so that a product may be manufactured economically. This will usually involve a knowledge of equipment, tool design, operation sequencing, worker/machine interaction, processing procedures, and facilities and plant layout. Manufacturing engineers will inevitably work with other industrial engineers engaged in plant design or in production and quality control.

PLANT DESIGN Engineers in this area will determine transportation systems for materials handling and conveyance and will specify the floor space needed for processing equipment and packaging. They will understand the economics governing plant location while designing for safe operation.

PRODUCTION AND QUALITY CONTROL The ultimate success and reputation of any industry often are governed by those working in the area of production and quality control. Industrial engineers must ensure that the end product meets certain performance standards by periodically checking its reliability through satistical studies, measurements, and other tests. A knowledge of production equipment and control components will be required so that production quality may be maintained.

DATA PROCESSING Many industrial engineers are involved in the enormous amount of data processing associated with the production of goods. Computers record inventories of materials and products along with many other management activities. Because they regulate many of the specialized processes required in production, computers are also quite important to the control of the production operation.

SYSTEMS ANALYSIS These techniques are applied by many industrial engineers to predict quantitatively the response of large industrial systems to the many variables that govern the operations necessary for successful production of goods. Operations research, optimization methods, linear programming, and queuing theory are mathematically based analysis techniques often applied by industrial engineers in systems analysis.

The majority of industrial engineers will be employed by manufacturing industries, but job opportunities will also exist in insurance and construction firms, mining companies, public utilities, hospitals, department stores, and federal and state agencies.

Industrial engineers are concerned with the management of industrial systems. This involves following a product from beginning to end, ensuring that the production processes and system management lead to finished products that meet quality standards. This industrial engineer is inspecting a finished product. (Courtesy of DuPont)

MATERIALS AND METALLURGICAL ENGINEERING

Technological innovation in the twentieth century has been made possible by timely advances in materials science and materials engineering. Scientists have investigated the physical or chemical behavior of solids, the origin of their properties, their response to varying environments, and their reaction to preparation and processing. By better understanding the principles governing strength, hardness, conductivity, magnetism, resistivity, and numerous other properties, materials engineers have been able to develop, manufacture, fabricate, and utilize improved materials in specific technologies.

Aerospace technology has provided a strong impetus to materials development. Space vehicles have required new fuels, and this naturally led to the

This huge fuselage section was bonded together with advanced adhesives developed by materials engineers. Adhesive bonding may be widely used in the future. (Courtesy of *Materials Engineering*, May 1980)

development of new materials to contain and control the chemical reactions and thermal dynamics of fuel combustion as well as extremes in temperature, acceleration, pressure, and radiation.

Vehicle control requirements have resulted in new and more compact electrical and mechanical devices to monitor and control displacement, attitude, vibration, and the thermal environment. In particular, we note the advances made in semiconductor materials. **Semiconductors,** such as silicon and germanium, have conduction characteristics that locate them somewhere between materials classified as conductors and materials classified as insulators. In the presence of certain impurities, their unique electrical conduction characteristics allow them to donate electrons (n type) or accept electrons (p type). They can be coupled in various ways to simulate the amplification and oscillation characteristics of vacuum tubes—at a fraction of the space and power. The correct proportioning of impurities has required new refining techniques such as zone refining. It has also led to a better understanding of crystal growth and to the development of materials having new optical qualities.

Whereas materials engineers are concerned with all solids—metals, ceramics, glass, and polymers—metallurgical engineers work only with metals. Since metals are crystalline solids, metallurgical engineers must be familiar with their microscopic characteristics. They will understand the properties of multiphase metals and their response to alloying and heat treatment. Some metallurgical engineers will develop new metals that provide greater strength, hardness, corrosion resistance, and endurance. Others may be concerned with maintaining strength at extreme temperature or with minimizing thermal expansion in metals. Still others may be interested in casting, machining, extruding, forging, and other fabrication and working techniques.

A major area of interest for metallurgical engineers is the extraction of metals from raw ores. This will necessitate an understanding of a variety of extraction and refining techniques such as zone refining, vacuum melting, and arc melting. Metallurgical engineers in this area will interact with other engineers interested in metals applications. Some may work in large metal-refining plants, whereas others may work with mining companies in the primary stages of recovering metals from ores.

Worldwide demand for resources should ensure a continuing need for materials and metallurgical engineers, although future engineers should expect to become increasingly concerned with materials recovery and materials recycling.

MECHANICAL ENGINEERING

"If it smells, it's chemical engineering!"
"If it sits still, it's civil engineering!"
"If you can't see it, it's electrical engineering!"
"If it is underground, it's mining engineering!"
"If it moves, it's mechanical engineering!"

From our previous descriptions of engineering fields, it is obvious that these somewhat facetious descriptions from the past are unrealistic, since there is considerable overlap from one field of engineering to the next. Yet there is some validity in the correlation of motion with mechanical engineering, because mechanical engineering deals with power, its generation, and its application; power requires the rate of change or "motion" of something.

Automobiles, engines, heating and air-conditioning systems, gas and steam turbines, air and space vehicles, trains, ships, servomechanisms, transmission mechanisms, and pumps are a few of the systems and devices requiring mechanical engineering knowledge. It is often argued that mechanical engineering is more broadly based than other fields of engineering, although this is becoming more debatable as other fields of engineering continue to diversify.

The major specialty areas of mechanical engineering are *applied mechanics; controls; design; engines and power plants; energy; fluids; lubrication; heating, ventilation, and air conditioning; materials, pressure vessels and piping;* and *transportation and aerospace.*

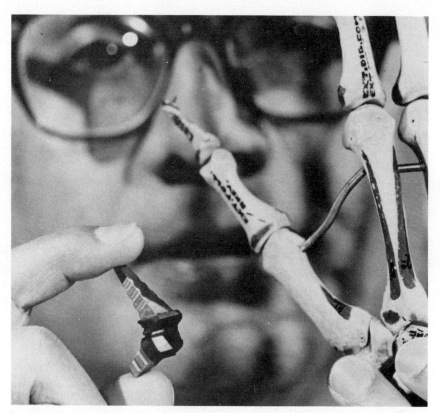

Progress is being made on materials for implants, such as this artificial finger joint, designed by engineers and made from titanium and a highly flexible rubber. (Courtesy *Materials Engineering,* May 1980)

APPLIED MECHANICS   Mechanics is that field of study concerned with motion and the causal effects of forces on this motion. Engineers in this area apply mechanics principles to the study, design, and development of systems and components that transmit specified motion, forces, power, or any combination of these. Mechanical engineers may study such mechanisms as cams and followers, may develop turbine shafts and blades that withstand the stresses and fatigue of higher rates of revolution, may develop structures that respond satisfactorily in shock and vibration environments, may build stability into orbiting satellites by spin stabilization, or may investigate stress/strain relationships for new materials.

CONTROLS   With the advent of the microprocessor, mechanical engineers in controls have entered a new era. The development of the large-scale integrated (LSI) chip has made compact, decentralized, digital, on-line data processing and control possible. Large amounts of data can now be compressed into small areas. When combined with the read-and-write capabilities of the microcircuit chip, data storage brings computer control into the reach of small as well as large companies. Controls engineers will take courses in control theory, mathematics, and systems. Because control systems often

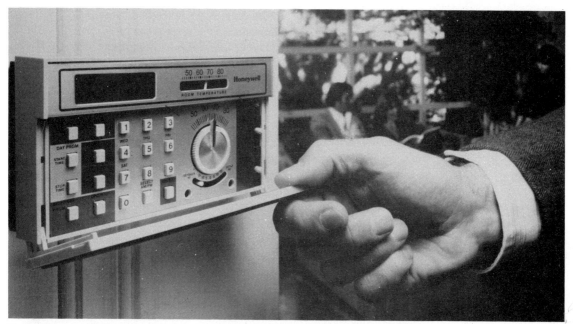

The problem of declining energy reserves has led engineers to apply advances in microelectronic technology to energy conservation. This thermostat stores day and night instructions to reset the desired temperature automatically, reducing fuel bills by 30 percent. (Courtesy of Honeywell)

include electrical, mechanical, and chemical components, it is not uncommon to find engineers from fields other than mechanical engineering working in this area.

DESIGN Mechanical engineers engaged in specialized design activities are numerous enough for us to call them **design engineers.** These engineers will have a working knowledge of materials and components and will understand the complexities and economics of assembling these components into products and systems.

ENGINES AND POWER PLANTS One of the greatest contributions of mechanical engineering is in the development and utilization of engines and power plants. Engineers in this area will work with reciprocating and rotating engines utilizing gas combustion or steam pressure to generate power that is transmitted through shaft motion. This shaft power may be used in power plants to generate electric power. It may be used to drive the wheels in ground transportation vehicles or to drive the propellers and turbine blades in airplane and jet engines.

ENERGY An important area for mechanical engineers is energy. Engineers are becoming more involved in designing solar, wind, geothermal, and nuclear-device systems and using them to generate power. This does not preclude their significant efforts in the more conventional power systems.

FLUIDS   Those who specialize in fluids are concerned with the flow of fluids, whether gases or liquids. They need to understand the basic characteristics of fluids and to distinguish them by their various properties, such as density, viscosity, and compressibility. Some engineers in this area investigate the properties of fluids in varying environments and the interaction of these fluids with solid bodies in motion; others develop new hydraulic control or power transmission devices.

LUBRICATION   Whenever machine parts move together—shafts and journals, gears, bearings, and so forth—they must be lubricated. Engineers will try to inhibit the wear on these parts by choosing or developing a lubricating method that minimizes friction and energy dissipation. This will require a knowledge of solid, liquid, and gas lubrication materials; lubrication methods and devices; and the surface wear and corrosion properties of a variety of materials. Lubrication engineers must understand methods for lubricant containment or protection and must be familiar with current methods for sealing, shielding, and filtering.

HEATING, VENTILATING, AND AIR CONDITIONING (HVAC)   The goal of engineers in HVAC is to control temperature and air quality. Current concerns include developing more energy-efficient equipment and systems and improving the integration of these systems into the design of structures in which temperature and air-quality control is necessary. HVAC engineers must understand basic heat transfer, thermodynamics, and control theory. They must also be familiar with systems and system components such as heat exchangers, compressors, expansion valves, thermocouples, and others that make up heating, ventilating, and air-conditioning equipment.

MATERIALS   Mechanical engineers also work with materials—their selection, development, and application. The selection and development of materials for bearings, brakes, clutches, gears, chains, screws, bolts, lubrication, insulation, heat transfer, and so on, will necessitate an understanding of material properties and the common test methods and equipment used to evaluate them.

PRESSURE VESSELS AND PIPING   Engineers in this area develop containment structures for solids, liquids, and gases that frequently must withstand high temperatures and pressures. They will have to understand the strength and flexibility characteristics of configurations such as plates and shells and the effect of joining them by rivets, welds, lamination, or other coupling means.

TRANSPORTATION AND AEROSPACE   Engineers in this specialty are engaged in the production or study of the motion of automobiles, trains, ships, planes, missiles, satellites, and rockets. Among their many responsibilities, they may develop improved gasoline or diesel engines, improve automobile power train transmission characteristics, modify the configuration of aircraft

Mining engineers are often concerned with the most efficient way of loading and moving coal from surface mining. (Courtesy of Ashland Chemical Company)

structures to minimize drag and increase lift, or improve solid propellant rocket engines.

Mechanical engineers, because of their diversity, require courses in heat transfer, thermodynamics, electricity and power generation, controls, vibration, fluids, design, and materials.

MINING ENGINEERING

Mining engineering is devoted to the discovery of metals or minerals found in ore bodies, to economic ore removal, and to the processing of these ores so that metallurgists can extract the valuable resources.

Modern technological means are employed in the discovery of resource solids; among them are aerial mapping, airborne magnetometers to search for irregularities in magnetic flux caused by resource deposits, Geiger counters to measure radiation, and earth resources satellites to scan the earth's surface for resources.

Actual removal of the raw resources falls into three categories: surface mining, underground mining, and ocean mining. Those involved in surface resource recovery, such as strip mining or quarrying, will have a knowledge of soils and rocks, blasting techniques, crack initiation and propagation, gravel management, and environmental restoration.

Underground mining engineers will be challenged to remove ores from below the surface economically and safely. The mining engineer may be required to understand the fundamentals of shaft sinking, blasting, boring, rock pressures, mine surveying, mine ventilation, hydraulics (including draining and pumping), mine timbering, power distribution, mine filling, subsidence, and mine management.

The continual depletion of continental resources is increasing the incentive to mine the oceans. Here, mining engineers are involved in the development of methods and equipment for both shallow and deep-water dredging. In the future, resources are likely to be recovered directly from seawater, which is high in dissolved minerals. Once the raw ores have been removed, the metals and minerals will be extracted by separation processes developed and applied by mining and metallurgical engineers. Techniques such as grinding and comminution (pulverizing) are followed by separation according to size, density, and magnetic, electrostatic, and surface-active properties.

Obviously, mining engineers must depend on a fundamental knowledge of geology and thus can expect to study and work with geologists. They will also interact with other engineers, particularly civil and metallurgical engineers.

NUCLEAR ENGINEERING

It is unfortunate that nuclear energy was first used destructively. Despite this heritage, nuclear engineering as a profession has continued to grow as new applications for nuclear energy are discovered. This growth has prompted the need for new engineering specialists who are schooled in atomic physics but interested in applications. The strong link with science caused most of the first nuclear engineering degrees to be offered at the graduate level and encouraged many pure scientists to pursue advanced degrees in nuclear engineering. In response to the increasing demand for nuclear engineers, colleges of engineering are now offering more degree programs in this discipline at the undergraduate level.

Among the largest employers of nuclear engineers are public utilities. Engineers here may be concerned with the design, development, construction, and operation of plants that utilize the heat produced by nuclear fission reactions (and, in the future, fusion) to generate electricity. In addition, nuclear engineers will have to solve problems associated with providing a safe working environment for plant personnel as they develop procedures for disposal of radioactive wastes and decontamination of radioactive areas.

Many nuclear engineering applications arise because of government-funded industrial research and development. We obviously can recognize the military and defense applications of nuclear engineering, particularly in our missile systems and in the nuclear power plants found in many of our submarines and ships. Less obvious are the applications of small nuclear plants to power satellites or to light navigation buoys or lighthouses.

Nuclear engineers have developed nuclear radiation equipment that measures the thicknesses of materials accurately, and this technology is currently being utilized in industrial mill-rolling operations and in quality control. Because of its depth-penetrating qualities, radiation can also be applied by engineers to measure the moisture content and composition of soils and cement.

Pickering Nuclear Power Station is located on Lake Ontario, east of Toronto. The site is dominated by the large vacuum building closest to the lake on the south. Four reactor buildings are joined to the vacuum building by a pressure relief duct. The turbine hall, auxiliary bay, and administration and service buildings are north of the reactor buildings. The station first fed power to the grid April 4, 1971, and was fully operational by June 1973. Total installed capacity is 2.16 million kilowatts. Construction on the "second phase" of the station began in 1974. (Courtesy of Ontario Hydro)

Among the many other applications developed by nuclear scientists and engineers are X-ray crack investigation; radiation modification of material properties such as strength, durability, and color; radiation sterilization and pasteurization; insect eradication; and food preservation.

Nuclear engineers will work with many other engineers and scientists in a frontier field and should expect to be continually challenged throughout their careers. In addition, they should be prepared to defend nuclear energy to those concerned about its use.

PETROLEUM ENGINEERING

Oil and gas have long been recognized as possessing unique energy-producing qualities. Before 1000 B.C., the Chinese were drilling for gas and oil and using it for lighting and heating. But it was not until the first important American well was drilled at Titusville, Pennsylvania, in August 1859, that a worldwide dependence on oil began. Within ten years, the United States was producing 4 million barrels of oil per year. Today, this country's annual oil consumption is over 10 billion barrels.

Obviously, the petroleum engineer has played an important role in the development of these energy resources by becoming involved in four major areas: locating petroleum resources, drilling for oil and gas, extracting these resources from the earth, and transporting and storing oil and gas.

LOCATING RESOURCES Locating and drilling for oil and gas has acquired worldwide significance in the major effort to find new reserves of oil. Besides the continental United States, major oil and gas fields are now located in and near Alaska, and in the North Sea, the eastern Mediterranean countries, Mexico, Central and South America, Canada, and the Soviet Union. Most of these have been established by American industries and have provided unique opportunities for petroleum engineers to travel the world and work hand-in-hand with geologists and geophysicists in geological exploration by seismic mapping and other subsurface techniques. Upon positive indication of gas and oil, petroleum engineers will initiate drilling in the hope of locating producible wells.

DRILLING FOR RESOURCES Petroleum engineers in drilling may be engaged in the design, development, and utilization of drilling rigs, platforms, and other equipment to drill thousands of feet into the earth's crust in the hope of reaching oil- and gas-bearing strata. They generally understand the mechanics of directional drilling and are able to control the hole pressures and drilling waste by circulating heavy muds through the drilling pipe. They also establish operational techniques to provide a safe working environment for construction and operational personnel.

EXTRACTING RESOURCES Although most drilling operations result in "dry holes," the occasional discovery of oil and gas means extraction or "production" must take place. Production engineers are involved in two important processes: precipitating the flow of oil within the reservoir to the pipe, and transporting it to the ground surface. In young wells the flow of oil and gas within the subsurface reservoir is caused naturally by down-hole pressure, and obtaining oil from such wells is called **primary recovery.** The present challenge to petroleum engineers is to recover the major portions of the oil and gas still remaining after the natural pressure has decayed. In these situations, petroleum engineers must apply such techniques for **secondary recovery** as water flooding, crack propagation, gas injection, emulsion flooding, in-situ combustion, and steam injection.

TRANSPORTING AND STORING RESOURCES As the sites where oil and gas are produced have become less accessible, petroleum engineers have become more involved in the design and development of methods and equipment to transport and store the product. Petroleum engineers in this area work with other engineers in developing land and water pipeline systems covering thousands of miles. Among the more notable ones are the aboveground Alaskan stainless steel pipeline with its environmentally protective heat-pipe support structures, the 1 000-mile pipeline from Texas to the Northeast states, the 1 400-mile gas line from Canada to California, and the North Sea underwater pipeline feeding several European countries. Where greater distances have been involved, petroleum engineers have assisted in the development of supertankers capable of carrying huge amounts of oil and liquified natural gas over thousands of miles. Resource storage, as the partner of transportation, has led petroleum engineers to develop new storage techniques found in the supertankers or in some of the unique floating offshore storage tanks in the North Sea.

The continuing depletion of our most accessible oil and gas is certain to challenge petroleum engineers to develop new methods to locate and extract oil and gas from areas previously unexplored or thought to be poor investments.

ENGINEERING TECHNOLOGY

In the late 1960s and early 1970s, a new member of the technological team emerged called the *engineering technologist.** Following a recommendation by the ECPD in 1971, several traditional two-year educational programs in industrial technology were expanded to four-year programs more in line with traditional engineering curricula, but less theoretical and more hardware oriented.

Since the initiation of these programs, thousands of technology graduates with bachelor of science in engineering technology (B.S.E.T.) or bachelor of engineering technology (B.E.T.) degrees have assumed industrial positions. Surveys have shown that the majority of engineering technologists have the title or job classification of engineer. These surveys have also shown that the starting salaries of engineering technology students approach those of graduating engineers, usually 5 to 10 percent less on the average.

Because engineering technologists are more applications oriented and less inclined toward the abstract innovation required in engineering research and development, the larger percentage of technologists are involved in design, computer programming, manufacturing, and production. This contrasts with the majority of engineers who are engaged in engineering research, development, design, and administration. Consistent with these trends are the educational differences at the graduate level between engineers and technologists. Surveys have shown that approximately one third of practicing engineers have completed an average of 1.5 years of graduate education, whereas less than one fifth of technology alumni have completed more than one year of graduate study.

---

*See Chapter 2 for a discussion of the technological team.

Characteristics of Technologists

Most of the characteristics exhibited by engineers will also characterize the engineering technologist. The major difference is that technology students become more familiar with common industrial practices and equipment than the engineering student, but less able to model and analyze physical situations.

Educational training of technologists will emphasize specific technical areas, leading to the development of specific skills that can be applied immediately upon employment. Thus, engineering technology students will specialize in a specific technology discipline and be introduced to the current practices within that discipline. For example, students in design technology would be trained in common design practices and, perhaps, trained to use computer-aided design (CAD) software to finalize the designs of engineers.

Engineering technologists seem to prefer more routine, standardized job environments, often applying their technical knowledge and skills to equipment, hardware components, maintenance procedures, and plant operations. Often their role will be one of support to other engineers or engineering development programs. Although they have been trained to use relatively current equipment and procedures, they retrain often to stay abreast of the new developments in their technical areas.

Engineering technologists will interact with managers, engineers, other technologists, technicians (many of whom they supervise), craftspeople, and equipment vendors and must therefore exhibit good communication skills. It is not unusual for those involved in plant operations to move into supervisory positions.

Technology Disciplines

A number of technical disciplines are available to technology students, depending on the breadth of technology programs offered in the local college or university. As can be seen from the following list, many of these disciplines parallel the engineering branches discussed previously in this chapter. Technologists in these disciplines will have knowledge and skills similar to engineers in the parallel branches but be more practice oriented.

Typical Technology Disciplines

Automotive technology	Manufacturing technology
Chemical technology	Mechanical technology
Civil technology	Mining technology
Electrical technology	Nuclear technology
Industrial technology	Petroleum technology

PROBLEMS

1.1 It is sometimes argued that engineering students would be better off graduating in one of the five major branches of engineering (chemical, civil, electrical, industrial, or mechanical), since these alone account for 80 percent of the graduating engineers. It is argued that these branches are more stable than the smaller branches. Discuss the pros and cons of this argument.

1.2 Write a two-page paper summarizing the important aspects of one of the following branches of engineering:

a. systems engineering
b. general engineering
c. marine/naval engineering
d. environmental engineering
e. geological engineering
f. ceramic engineering

1.3 Interview two engineers from a branch of engineering of your choice, preferably the one that you have chosen for your engineering career. Write down the similarities and differences in their perception of this branch of engineering.

1.4 List the important specialty areas for the branches of engineering that are described within this chapter. Into what specialty areas might the engineers you interviewed in problem 1.3 fit?

1.5 Interview an engineering technologist and write down that person's perception of the differences between his or her job responsibilities and those of the engineers with whom the technologist works.

1.6 Investigate the origins of one of the following branches of engineering:

a. ceramic engineering
b. marine/naval engineering
c. nuclear engineering
d. computer engineering
e. industrial engineering
f. aerospace engineering

1.7 What new branches of engineering do you foresee coming into existence before the year 2000? Why?

1.8 Investigate the engineering job opportunities available to the graduating seniors at your institution.

1.9 During the late 1970s and early 1980s, the country was in a recessionary period with higher unemployment than normal. Yet, the demand for engineers continued to increase. Explain.

1.10 Write a short essay tracing the contributions of electrical engineers to the rapid growth in microelectronics.

1.11 Why has the modern definition of engineering evolved to include the phrase "with concern for the environment and society"? Discuss at least five technological innovations of the twentieth century that have had unfavorable side effects.

2 Design and the Engineer's Function

CONTENTS

Engineering Design

The Design Process

The Phases of Design

The Technological Team

Engineering Functions

Research: Knowledge, Understanding, Application

Development: Methods, Devices, Performance

Design: Selection, Specification

Production and Construction: Processes, Assembly, Structures, Systems

Operations and Maintenance: Performance, Service, Repair

Sales: Marketing, Applications, Service

Management: People, Projects, Products

Looking Back

Most engineering activities relate in some way to the larger problem-solving activity called engineering design. What role the engineer plays in the overall design process is defined by the engineering functions.

ENGINEERING DESIGN

Engineering design is a decision-making or problem-solving activity that engages not only engineers but technologists, scientists, technicians, craftspeople, managers, and many others. This cadre of specialists, sometimes called the **technological team,** is discussed at greater depth later in this chapter. The ultimate purpose of design is to convert resources into devices, systems, processes, and products to meet human needs. Engineering design can be arranged schematically to display the "iterative" process characteristic of the more difficult design problems (Figure 2.1).

Design iterations are necessary because engineering design strives for optimal solutions that satisfy the design requirements and any design restrictions (also called design constraints). When a solution fails to meet either the restrictions or requirements placed on the design, it is necessary to modify the solution or even to look for an entirely new solution, that is, a design iteration. Most iterations will occur within the block designated in Figure 2.1 as **primary design,** sometimes referred to as engineering design.

In engineering, designs are conceptualized, developed, built, tested, evaluated, and then redesigned to provide near-optimal solutions that satisfy the requirements and restrictions that are imposed on the design. Engineers of various specialties integrate their efforts with those of other technical specialists to develop plans and specifications from which the final device, product, or system can be built or produced. It is not unusual for the most numerous and significant design changes or iterations to occur during this stage; we refer to these as the **primary iterations.**

In reality, engineering design permeates all of the steps shown in Figure 2.1, simply because engineers are engaged in activities other than those referred to earlier as primary design. These activities can be placed into certain engineering functions, as described later in this chapter. As we will

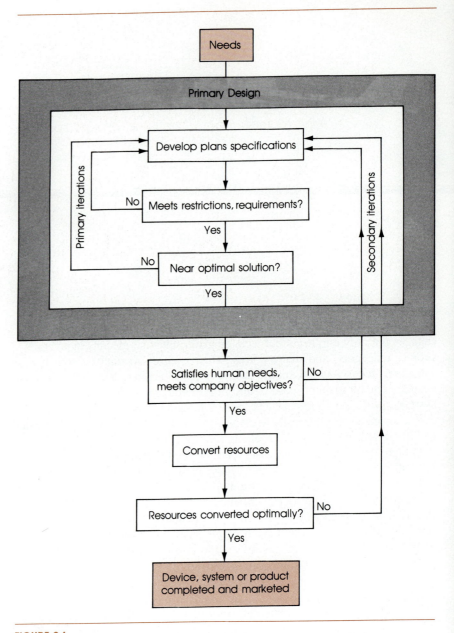

FIGURE 2.1

Engineering design.

learn, the engineering functions describe the engineer's part in the design evolution of some new process, project, or product.

Although most design changes occur during primary design, the need for some changes may not become obvious until after construction or manufacturing operations have begun or after company management scrutinizes the

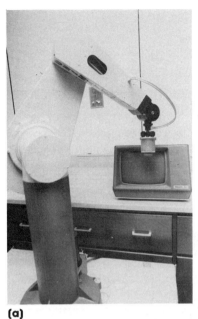  

(a)　　　　　　　　　**(b)**　　　　　　　　　**(c)**

Engineers will take a new technology like the robot shown in (**a**) and first determine performance in the laboratory, sometimes using complicated models like the sophisticated erector set shown in (**b**). Finally, engineers will integrate this technology into the industrial environment (**c**). This process usually takes time and involves a number of engineers having different expertise. (Courtesy of Unimation and Texas A&M University)

design—or even after the design has been constructed or marketed. In general, though, these design iterations are of a secondary nature. Those design flaws discovered after the device, product, or system has been completed and marketed are usually attributed to a use environment for which design criteria were not identified. This has been the case in the modern automobile, where design flaws have been discovered in steering mechanisms, fuel systems, braking systems, and transmissions, or where seemingly unpredictable human-engineering flaws seem to surface only after the car has been marketed in quantity. Design changes that are made after the product has been marketed—through product recalls and retrofit—seriously reduce profits, and most industries employ extensive testing procedures to minimize these kinds of design errors.

Design Education

Undergraduate engineering students usually engage in design peripherally, with most of their time spent in problem solving—an important activity that is the basis of the engineering design activity. Students solve reasonably well-defined problems that are less open-ended than those faced by practicing engineers, who often must reconcile design objectives, restrictions, and other criteria that seem to conflict. This is not to imply that engineering students receive no training in real design. In fact, most upper-level students will take design courses in which the design problems are less defined than those at lower levels. The purpose of these courses is to teach students the advanced

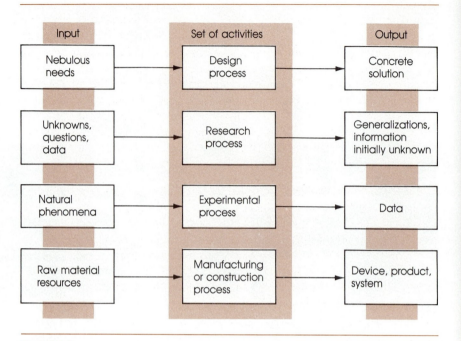

FIGURE 2.2

Input/output processes.

problem-solving skills required in real design by having them struggle with problems in which several solutions may be possible.

Because of their lack of experience, students should not expect, upon graduation, to assume major roles in the broader problem-solving activities of design. Nevertheless, college work will demonstrate the challenges involved in the industrial problem solving we call engineering design. This chapter, it is hoped, will make both engineering and technology students more cognizant of the differences between the college and professional environments and cause them to approach their problem-solving and learning activities from a broader perspective.

The Design Problem

In engineering design the expressions *process of design* and *phases of design* are often encountered during a discussion of design morphology (the steps or structure by which design solutions evolve). Although similar in many ways, they cannot always be used interchangeably. The process of design is an ordered method of attacking design problems or their subproblems. It is similar in many ways to the structured problem solving that engineering and technology students engage in, but it is not restricted to finding the "unique" solution often required of students. As shown in Figure 2.2, it can be envisioned as a set of activities that translate a design from its nebulous initial or input state to an output state characterized by some concrete solution.

The design process can be compared to the research process, another problem-solving activity, the objective of which is to move from an initial state characterized by unknowns, questions, and data to a desired final state in which new knowledge is uncovered and where generalizations can be made that were initially unknown. It can also be compared to the manufacturing or construction process, which converts input resources to output devices, products, or systems, or to the experimental process by which output data are obtained from the natural phenomena that initiate the experiment. These processes are similar because each deals with discovery and implementation, either in abstract equations, data, or graphs or in tangible systems, products, and devices. As an example of this discovery process, consider the following design problem.

THE PROBLEM   A variable-speed pulley used in torque converters transfers torque to the shaft through a hub/shaft, keyway, and key arrangement (Figure 2.3**a**). The axial movement of the pulley discs, designed to occur as the pulley force varies, subjected conventional steel keys and their rectangular keyways to binding, fretting fatigue, and fretting corrosion (fretting is local wearing and corrosion caused by high contact forces over localized areas of contact) and to failure by shearing across the cross-sectional area of the key. The objective was to correct this problem without significantly changing the basic design of the pulley and shaft system.

THE SOLUTION   Engineers recently solved this design problem by modifying the previous design approach to arrive at a solution that was simple, yet satisfied all the design criteria. The solution (Figure 2.3**c**), named the Quadra-Key, overcame the conventional key's problem of concentrated loading due to cocking (Figure 2.3**b**) by essentially rotating the rectangular key and keyway 45 degrees and then flattening two opposing corners. This modification makes it difficult for the key to cock and causes the torque loads to be distributed over a larger contact area between the keyway and key. In addition, the shear area is increased because the shearing area across the diagonal of the key is 40 percent greater than the shearing area of the conventional key. Additional advantages were gained by manufacturing the keys from a proprietary nonmetallic material that is self-lubricating and resistant to fretting, corrosion, and binding.

Negative Design Decisions

Not all designs like the Quadra-Key are successful. The conclusions at the end of each phase may not warrant continuing the design. A recent example is the Boeing supersonic jet transport (SST). Even though the design had progressed far enough to construct a functioning prototype, environmental, economic, and technological studies concluded that the time was not right for a supersonic transport. Despite these conclusions, the French and British built and implemented their supersonic transport, the Concorde, and the Russians built theirs, the TU 144. Because of high operating costs and significant technological problems that often grounded them, both were financially unsuccessful. Obviously, these cases point out the need for effective design evaluation; negative design decisions should be made early in the design,

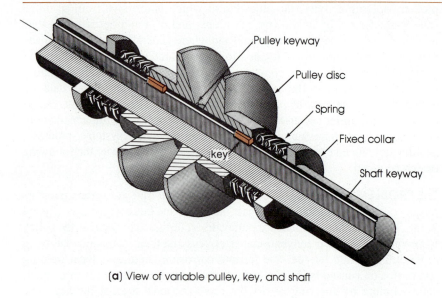

(**a**) View of variable pulley, key, and shaft

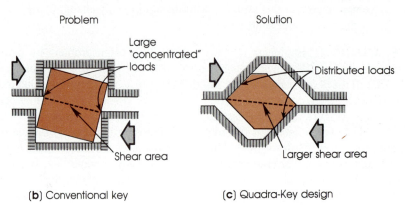

Problem

Solution

(**b**) Conventional key

(**c**) Quadra-Key design

FIGURE 2.3
Design of an improved key.

preferably in the first phase, when feasibility is investigated and minimum company resources have been committed.

THE DESIGN PROCESS

The design process is a method of problem solving. Engineers engaged in problem-solving activities, particularly engineering designers, should have or develop the following characteristics and abilities:

- **Knowledge** of physical laws, mathematical principles, computers, materials, and the common engineering applications, whether in methods, projects, products, or processes.

- **Skills** of analysis, communication, and graphical expression.
- **Attitudes** that include motivation, curiosity, intellectual and personal integrity, confidence, flexibility of thought, patience, and willingness to take risks.
- **Creativity.**

Engineers with these characteristics and abilities are better prepared to solve the numerous problems that normally arise during design. Nevertheless, knowledge, skills, attitudes, and a creative nature do not guarantee success in solving problems; they just provide the necessary ingredients that, when combined with hard work, produce a desired result.

Unfortunately, there is no formula that engineers can apply to all design problems. Each design project has its own personality. This is why the steps stated for the design process often vary. Differences lie in where to begin and end the design process. Some feel that the design process should be applied after the problem is identified; others, that the design process should begin in the initial problem identification step. Then there are those who feel that the design process ends with design specification, while others would see it continued to implementation, and even through distribution and consumption. The eight steps listed in Box 2.1 reflect the opinion that the design process never ends. Design iterations occur after the design is *completed*— even after the final design is marketed. Most current models are usually revised to accommodate newer devices, equipment, systems, and materials in an ongoing process of design. As mentioned earlier, design flaws still arise in marketed products, necessitating further design modification.

Box 2.1
The Design Process

Step 1: Problem identification
Step 2: Gathering of information
Step 3: Preliminary solution ideas
Step 4: Modeling, analysis
Step 5: Testing
Step 6: Decision
Step 7: Specification
Step 8: Implementation

THE PHASES OF DESIGN

There is little difference between the phases of design and the design process applied to the overall design. The major difference is that the phases of design break the *overall* design into several chronological *stages* of completion. In

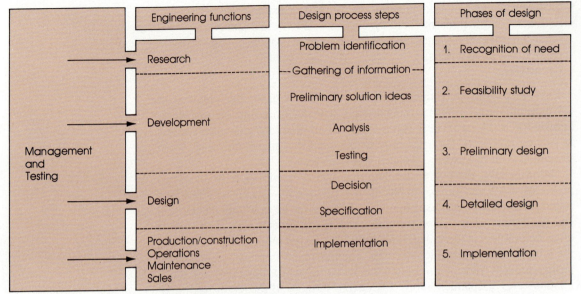

FIGURE 2.4

Chronological comparison between the engineering functions, the design process, and the phases of design.

contrast, the design process may be applied not only to the overall problem but to any of the associated subproblems or subdesigns.

When the design process is applied to the overall design project, it follows a chronological pattern similar to that in the design phases. In fact, even the engineering work *functions,* when placed in a certain order, fall into a similar chronological pattern, with research and development initiating the process (Figure 2.4).

Following the initial **recognition of the need** phase, we enter the final four phases of design: the **feasibility study,** the **preliminary design,** the **detailed design,** and the **implementation** phase. We identify the main activities of each phase so that students may see how chronological progress is measured in design (see Box 2.2).

Box 2.2
The Five Phases of Design

Phase 1: Recognition of Need
Go to Phase 2.

Phase 2: Feasibility Study
1. Define the elements of the problem. Identify all the design requirements.
2. Identify the factors or restrictions that may limit the scope of the design.

3. Identify any difficulties that might arise owing to competitive products, unusual use situations, unexpected environments, cash flow difficulties, or incompatibilities.
4. Consider possible consequences of the design.
5. Generate alternative solutions.
6. Evaluate the alternative solutions for feasibility.

If one or more of the alternative solutions appear feasible, go to Phase 3. Otherwise, start over or give up.

Phase 3: Preliminary Design
1. Select the best solution(s) as they now appear, either one design solution or several.
2. Consider all parts, components, materials, processes, methods, strengths, and dimensions for each of the possible design solutions.
3. Consider the availability of materials, components, labor, buildings, land, and finances relative to each of the candidate solutions.
4. Determine the impact of each design solution on society and on the natural environment.
5. Consider the impact of the environment on each design. This includes the operational environment, the marketplace environment, and the manufacturing-plant/construction-site environment.
6. Model, analyze, and test each candidate design as necessary.
7. Evaluate and compare the candidate design as objectively as possible. Determine the best solution.

If the best solution appears satisfactory, go to Phase 4. Otherwise, start over, modify, or give up.

Phase 4: Detailed Design
1. Examine all principal requirements and restrictions in depth. Consider abnormal situations.
2. Refine models and analyze them in depth.
3. Test for strength, endurance, and life and correlate test results with analytical results.
4. Specify the design in detail, coordinating the details with those involved in production, construction, and marketing. Economics becomes a major factor in determining the components and materials used and their availability.
5. Record the design details in reports, drawings, graphs, and charts.

Continued on next page

If the detailed solution is satisfactory to the decision-makers, go to Phase 5. Otherwise, modify, start over, or give up.

Phase 5: Implementation
1. Identify the materials, personnel, processes, methods, marketing, financing, and facilities that are required.
2. Integrate the materials, personnel, processes, methods, marketing, financing, and facilities into a functioning organization.
3. Produce or construct the design in an optimum manner. Monitor the economics of production, the production rates, and the production or construction quality.
4. Market the product or sell the system.
5. Monitor the success of the product or system in the marketplace and make design modifications as necessary.
6. Recover resources as necessary. In other words, recover things like used bottles, scrap metal, and paper.

If the design is profitable, continue production, construction, and marketing. Otherwise, modify, redesign, or abandon the design.

THE TECHNOLOGICAL TEAM

Although engineers play a major role in engineering design, they are just one contributing component of a technological team composed of mathematicians, scientists, engineers, technologists, technicians, managers, and craftspeople. On an ordered scale, the top of the scale indicating interest in abstract concepts and theoretical principles and the bottom indicating interest in or capacity for performing specialized manual skills, often of a technical nature, we would expect mathematicians and scientists (including computer scientists) to be at the top, technicians and craftspeople to be at the bottom, and engineers, managers, and technologists to be somewhere in the middle (Box 2.3). Although, strictly speaking, mathematicians are scientists, we have distinguished them from those scientists who deal with more physical things, for example, physicists and chemists.

Educational background and attainment also distinguish the members of the technological team. Mathematicians, scientists, and engineers usually have a bachelor of science degree (B.S., four years of study) as a minimum and often have obtained master of science (M.S., five to six years of study) and possibly doctor of philosophy degrees (Ph.D., eight years of study) at institutions of higher learning.

In contrast, managers may have either business or engineering educational backgrounds with respective bachelor of arts (B.A.) and B.S. degrees. Many

Box 2.3
The Technological Team

Abstract Concepts, Theoretical Principles

Mathematician Investigates quantities, magnitudes, and forms and relates their attributes by the use of numbers and symbols.

Scientist Seeks to understand nature and the universe by applying observation, study, and experimentation and by developing principles and methods.

Engineer Applies a knowledge of mathematics and physical science to the creative, economical solution of problems using engineering principles and methods.

Manager Supervises and controls the personnel and resources required for the development and design of new technological products or processes.

Technologist Implements designs and develops the ideas of scientists and engineers for production, construction, and operation, solving many of the practical problems that arise in the process.

Technician Applies specialized, less conceptual skills to assist scientists, engineers, and technologists to implement their plans and designs, often assembling, operating, and maintaining equipment, preparing drawings, and estimating costs.

Craftsperson Applies specialized manual skills requiring physical dexterity to produce products or construct larger structures such as roads, plants, and buildings.

Specialized Manual Skills

INCREASING ABSTRACTION

have upper-level degrees such as the M.S. or the master of business administration (M.B.A.) degree requiring one or two years of additional study beyond the bachelor's level.

ENGINEERING TECHNOLOGIST The newest member of the technological team, the technologist, requires four years of education in an engineering technology (E.T.) program, leading to the baccalaureate degree commonly designated the bachelor of science in engineering technology (B.S.E.T.) or bachelor of engineering technology (B.E.T.) degree. The educational curriculum will emphasize science and mathematics along with the more practical applications of technical principles, processes, and equipment. Note that because engineering and E.T. students take many similar courses during the first two years of study, a transfer from an engineering curriculum to an E.T. curriculum or vice versa can be made without a significant loss in curriculum course work.

ENGINEERING TECHNICIAN Usually the technician will require two years of college-level study leading to the associate degree in engineering technology, taken at a technical institute or community college. This educational program emphasizes practical techniques and a thorough familiarization with the operation, limitations, and maintenance of technical equipment. Much of the course work would be transferable to engineering technology programs but not to programs leading to degrees in engineering and science.

CRAFTSPERSON Craftspeople like plumbers, electricians, machinists, and carpenters practice manual skills that are learned by experience, often requiring an apprenticeship period directed by experienced craftspeople that may range from six months on up. A high school education is an expected prerequisite.

Every member of the technological team is important to the progress of technology, and individuals are deriving satisfaction performing each of the functions described in Box 2.3. Students must carefully assess their interests and abilities before selecting a career. Recognize that salaries are not necessarily commensurate with the degree of education or the relative ordering in Box 2.3; rather, they reflect the level of responsibility for the work of others. Thus, managers usually make the higher salaries.

A more detailed explanation of what engineers do and how they relate to the other members of the technological team follows.

ENGINEERING FUNCTIONS

There is a difference between "what" engineers do and the branch of engineering and specialty area "where" they do it. The "what" can be one or more of several engineering functions or activities necessary to the successful completion of any engineering task. Those listed below and described in the following sections involve most of the engineers and some technologists who are engaged in project or product engineering.

- Research
- Development
- Design
- Production and construction
- Operations and maintenance
- Sales
- Management

All these functions represent important steps in the design evolution of a new or improved product. They do not segregate engineers and technologists by branches but by interests, capabilities, and experience. In fact, young engineers should expect to perform one or more of these functions during their professional careers, although most will move toward one fixed work function as time progresses.

Consulting, Teaching, and Testing

Although they are important engineering functions, consulting, teaching, and testing are not included in the foregoing list because few engineers are engaged in them. Consulting engineers perform engineering services for the government or industrial companies, but usually on a temporary basis. They offer specialized knowledge, skills, and experience that may not ordinarily be found among the technical capabilities of companies. Since it is much cheaper to hire a temporary "expert" than to employ a full-time specialist, this can prove quite cost effective for companies that sporadically need specialized technical services.

Among engineers who act as **consultants,** it is not unusual to find engineering teachers consulting part-time, while full-time consultants are more apt to be self-employed or engaged with others in a small consulting firm. Engineers who offer consulting services, particularly on a full-time basis, are running a business that may serve local or federal government or private industry. They will need to understand business management principles, understand the legal responsibilities of their engineering decisions, and be registered professional engineers.

Some engineers enjoy the learning environment and pursue their careers as **engineering teachers.** Because they are engaged in the transfer of knowledge, most will obtain advanced engineering degrees and will continue to expand their capabilities through further study, research, consulting, sabbaticals, or summer employment in industry.

Perhaps the most important contribution that an engineering educator makes comes in the one-to-one interaction between students and teachers. Through their explanation of problem solutions, engineering teachers encourage students to develop their problem-solving abilities and their understanding of the engineering design process. In addition, this interaction between teacher and student gives many students their first exposure to the professional aspects that underlie engineering decisions.

In many cases **engineering testing** takes on the role of a separate engineering function supporting almost all the functions that will be discussed in the following sections. Larger organizations frequently have separate departments of test engineers who have the major responsibility for conducting and managing the test programs for other company groups and departments. This may involve the selection of test materials and equipment, scheduling of facilities, production of specialized parts, and the direction of technicians and technologists who may actually conduct the test. Test engineers usually gravitate to this function because of their interest in working with physical objects in an environment of evaluation and assessment.

Projects, Processes, Products

Before launching into the description of the various engineering functions, it is necessary to briefly examine the similarities and differences between engineering projects, processes, and products.

Projects are usually activities of long duration and low volume. We speak of development projects, construction projects, or production projects. Whenever an organization initiates engineering activities that have certain specified goals—such as a new bridge, a new structural analysis computer program, or a new chemical process—it usually becomes "project oriented," requiring schemes for project organization and progress assessment. Invariably, the overall project will be broken into *mini* projects, each crucial to the completion of the primary project goals.

Processes are the means by which materials are made or transformed into some useful item called a product. Processes may lend themselves to mass production, in which high volume is the object, or they may be employed within a project. In some cases a new process may be the object of a company project. In the sense that a **product** is something made by industry, even dams and ships are products and thus represent the completed project goal. Usually, products are thought of as smaller items produced in volume.

RESEARCH: KNOWLEDGE, UNDERSTANDING, APPLICATION

The aim of **basic research** is to establish new principles or to explain scientific phenomena previously not understood; in other words, to expand knowledge. In contrast, **applied research** is concerned with the constructive application of this new knowledge. Thus, research engineers are usually engaged in applied research, while scientists and mathematicians are more apt to be involved in basic research, although there can be considerable overlap.

Rarely will engineering technologists lead research activities. Nevertheless, they may work with engineers on a research project.

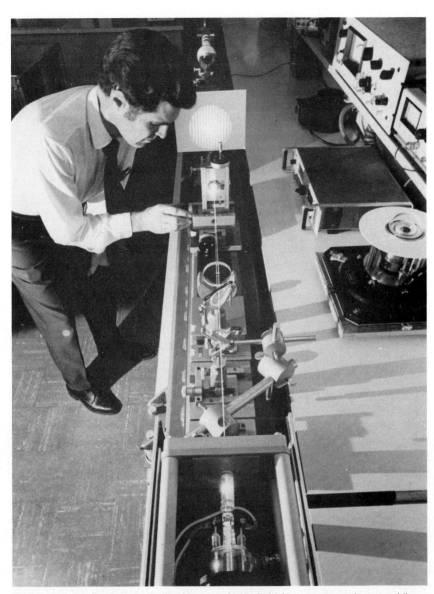

This experimental apparatus splits a beam of blue light from an argon laser and then crosses the two beams to develop interference fringes. Fringe systems are the basis of a new optical method for measurement of fluid velocity known as laser doppler velocimetry. Here the engineer is using a traveling microscope to measure the fringe spacing. (Courtesy of Pratt and Whitney Aircraft Group)

Research engineers search for the "if we can," usually in response to an immediate need for a new technological breakthrough. For example, the Apollo, Viking, and Space Shuttle programs required significant research efforts to push the technological state of the art forward. New metals, paints, and ceramics were developed in response to the need for materials more resistant to thermal extremes. But these resulted only after extensive engineering and scientific research into the nature of materials. Research breakthroughs into the properties of semiconductors that occurred in the late 1940s eventually allowed aerospace engineers to develop the more sophisticated microcircuitry and minicomputer components needed.

Other research provided synthetic composites formed by integrating the desirable properties of individual materials into a new material possessing all the desired properties. One example of a widely used composite material is fiberglass. Combining the strength of pure glass fibers and the flexibility of plastics, research engineers carefully oriented the fibers in a plastic-based matrix to derive a composite material stronger than steel on a strength-per-density basis. Fiberglass has replaced the steel in fishing rods and the wood and metal in many boat hulls and is used extensively where shell structures are required, for example, swimming pools, water and fuel tanks, and automobile bodies.

Research is needed not only to support large programs such as the Space Shuttle, but in all industries that are developing new products and equipment. Automobile companies support research that may provide new engines having greater thermodynamic efficiencies or lighter transmissions and body structures that improve automobile mileage; computer industries are supporting research into the further miniaturization of electronic components; and energy industries are sponsoring research to find new methods of secondary recovery of oil and to develop the economical means of utilizing the energy contained in our geothermal resources and oil shale.

Qualifications

What, then, are the qualifications that an engineer must meet if he or she wants to do research? Among them are superior intellectual ability with an education beyond the undergraduate level, the desire to tackle poorly defined areas of research with patience and openmindedness, the ability to understand basic scientific and engineering principles and to apply them in innovative and creative ways, and the ability to work and communicate with others in cooperative research ventures. Of all the engineers with doctorate degrees, approximately half function as research engineers, whereas only 15 percent with master's degrees perform research. The percentage drops even further— to 5 percent—at the bachelor's degree level.

Opportunities

Opportunities for engineers to perform research have increased considerably in the last two decades. Major industry today cannot afford to neglect research if it is to remain competitive. This becomes readily apparent when the total $1 million expenditure by the 100-odd research labs operating in 1920 is compared with the billions spent by thousands of research labs today.

Research laboratories and organizations today either are appendages of large companies or are supported by the government and universities (those supported by universities are often called research institutes). Research is not always confined to large laboratories or organizations but is often integrated with the other major work functions, usually on a reduced scale. For example, university and college teachers may pursue funded research on a part-time basis.

DEVELOPMENT: METHODS, DEVICES, PERFORMANCE

Conducting research without planning for development is the same as nurturing a fruit tree that holds no promise of bearing fruit. This is because the natural purpose of development is fulfilled only when the new ideas and concepts of research scientists and engineers are applied and utilized in some useful way. It is this intimate relationship between research and development that has led to the term R&D.

Qualifications

It is not uncommon to find research and development engineers alternating between these two functions or at least closely monitoring the activities of both. In fact, most development activities are preceded by an investigation or survey to determine what new knowledge or new research discoveries are available. Thus, it is not surprising to find engineers with advanced degrees not only in research but also in development. Approximately 24 percent of the engineers having doctorate degrees and 28 percent of those with master's degrees are engaged in development work. Twenty percent of all engineers work in engineering development, and the majority of them have bachelor's degrees. These numbers lead us to the obvious conclusion that development work is truly one of our major engineering activities, requiring a significant commitment of personnel and resources. It seemingly has the capacity to absorb all kinds of engineers and promises an interesting and dynamic environment. Development will continue to be an important work function for engineers as long as our society promotes technical innovation.

Example: The Heat Pump

One interesting example of development failure and success is the heat pump, a product of the early 1950s. Development engineers had responded to the need for more efficient, more versatile, electrically run heating and cooling systems by developing the reversible heat pump. Within a short time there were 43 different brands of heat pumps on the market. Almost without exception, however, they proved to be unreliable because their design depended on air-conditioning components, and these had not been developed to withstand the more severe environment in which heat pumps are used.

A heat pump is essentially an oversized refrigerator. In the refrigerator, the aim is to cool a compartment by removing excess thermal energy that is then rejected as waste heat. This is done by circulating a fluid refrigerant, usually

(a)   Heat pump cooling

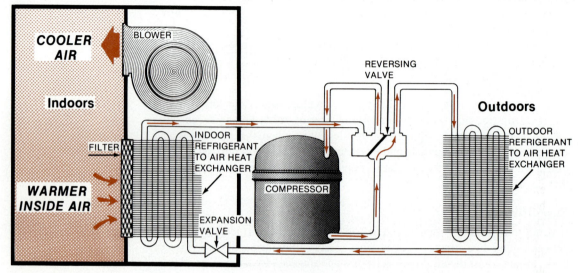

(b)   Heat pump heating

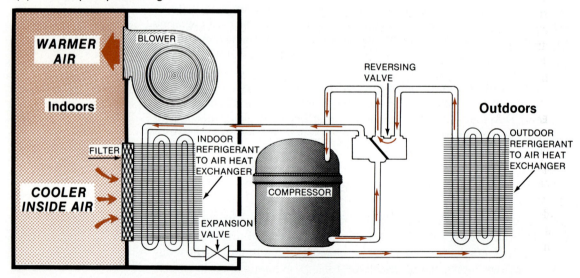

FIGURE 2.5

Heat pump cooling and heating cycles.

Freon, between the evaporator coils in the area to be cooled and the condenser coils where waste heat is to be given off (Figure 2.5).

The overall cycle begins as the refrigerant leaves the evaporator as a low-temperature, low-pressure vapor and enters an electrically driven compressor, where it is compressed to a hot, dense vapor before entering the condenser. By giving off heat to the kitchen at lower temperature, the high-pressure vapor condenses to a high-pressure, cooled liquid, which then flows to a restriction device called an expansion valve. This allows the high-pressure liquid to expand and vaporize, causing the Freon to chill to a temperature below that of the refrigerated compartment as it enters the evaporator coils. It then can absorb any excess heat in the compartment and begin the thermal cycle again.

When the heat pump is used as an air conditioner during the summer, the "compartment" to be refrigerated is simply the living area. The excess heat is rejected outdoors. During the winter, the heat pump is reversed to draw heat from the cold outdoors and then "reject" it in the condenser coils to the living area. Obviously, both the cooling and heating modes require a fan-driven air distribution system.

The heat pump's versatility now becomes apparent. It can function as an air conditioner in the summer and a heater in the winter. Thus, a consolidated heat pump system offers space advantages over separate heating and air-conditioning systems. If the separate heating system is a less-efficient electric resistance device, the heat pump offers cost advantages as well. When one considers that about a sixth of our total energy and a fourth of our oil and natural gas are consumed in environmental temperature control, the advantages of the heat pump become even more significant. Whereas electric resistance heating brings heat into the heated area by electrical transmission over great distances, the heat pump requires only one half or one third of the electricity to accomplish the same thing by simply "shifting" some of the outdoor thermal energy indoors for heating. These advantages prompted the proliferation of heat pumps.

THE EARLY MISTAKES

What the first heat pump engineers failed to recognize was that the earlier air-conditioning-type compressors were not satisfactory for the lower-temperature environments characterizing the heating cycle. The lower temperatures reduced the Freon vapor pressures, thereby placing greater stress on the compressor. In addition, the withdrawal of heat from an already cold outdoor atmosphere caused ice to form on the outdoor coils, rendering the heat transfer process ineffective and thereby placing additional strain on the compressor.

THE IMPORTANCE OF TESTING

Where did the development of these early heat pumps fail? Without doubt, the failure occurred because of insufficient testing, the most important and culminating activity in the development of products. It is in testing that material, structural, and operational performance is examined and serious flaws are uncovered. Field testing is particularly important when a system or product such as a heat pump must operate in an environment not easily duplicated in the laboratory.

Although the early heat pumps had serious design deficiencies, engineers later remedied the problems by developing heavier-duty, higher-efficiency compressors that were lighter, were hermetically sealed against environmental infringement, and had improved bearings, seals, and other materials. These developments have reduced overall compressor failures over the warranty period to less than 5 percent. Thermostatically controlled, quick-"defrosting" cycles moderated many of the icing problems, and the electrical complexity of the units was reduced significantly by modern electronic and control components. In addition, the heat exchange process was facilitated by adding fins and other advanced geometric appendages to the outdoor coils to increase the heat exchange surface area.

Despite these improvements, overall heat pump performance is still a fraction of that theoretically possible. The two to three values presently obtainable for the coefficient of performance (COP) are far below the eight to ten theoretically possible. And although it is unrealistic to believe that future developments in heat pump technology will move COP values much above six, significant room for development progress remains.

Development Steps

Most development projects go through several distinct development steps. The descriptions in the following paragraphs include the appropriate qualifications engineers must bring to each step.

INITIATION OR "SELLING" OF A PROJECT The initial step in development usually comes in response to research discoveries, in response to marketing results that indicate some need, or because an engineer, inventor, or group of engineers successfully "sells" an idea to management by promising a return on investment. Development engineers here must be imaginative and able to sell ideas.

UNDERSTANDING THE PROBLEM Here the engineer carefully reviews the constraints imposed on the project. These constraints may be monetary, as with a less expensive calculator, or they may be physical, requiring, for example, horsepower for less weight or improved electrical conductivity at higher temperature. Experienced development engineers will quickly recognize those constraints most difficult to meet, because they are familiar with other similar systems or components. In addition, the development engineer will search the literature for the latest information, equipment, and developments that may prove helpful. Engineers here must be well educated and experienced; they must be able to accept a certain amount of uncertainty and to distinguish important information from the unimportant. Familiarity with the state of the art in new technologies is critical, requiring engineers to continually study technical literature and new innovations. (See Chapter 7 for an introduction to the technical library.)

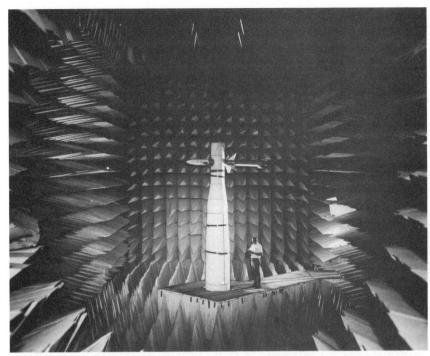

This anechoic chamber is used for the accurate measurement of sound. (Courtesy of U.S. Navy)

ALTERNATIVE SOLUTIONS   The sole purpose of this creative step is to generate a number of feasible solutions that appear to meet the specifications and constraints set for the project at hand. It is a mental process requiring the engineer to conceptually create one or more new devices or methods to produce the desired result. Often the engineer leans on the process of synthesis in this step. As a solution begins to take shape mentally, it is common for rough models to be constructed so that the step between abstract concepts and tangible objects might be made.

DEVELOPMENT MODEL(S) AND TESTING   On the basis of the results from the preceding step, one or more models will be constructed, usually by hand, that appear to meet the requirements initially imposed on the development project. The model may include new, one-of-a-kind components or off-the-shelf items placed in some new and unique arrangement. Testing on the model or models will verify predicted performance data or will expose design problem areas that will necessitate model revisions. Technologists, technicians, and craftspeople will usually do the construction or run the tests under the careful supervision of development engineers or specialized test engineers. But it is the responsibility of the engineer to analyze and interpret the test result accurately, a crucial step to overall project success. From these observations will come the engineer's recommendation to management to move full speed ahead (necessitating detailed design and production), to

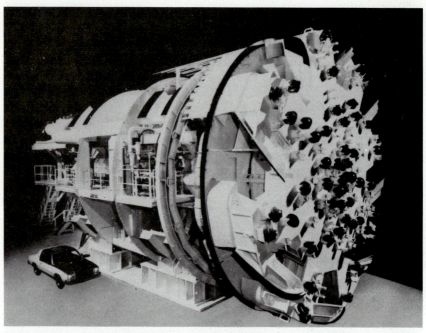

Some engineering designs are quite complicated. This tunnel-boring machine (TBM) built by Jarva, Inc., can bore a 32-foot tunnel through solid rock at speeds up to 80 feet per day. Only one person is required to operate the TBM and to keep it on line and grade. (Courtesy of Foster, McGuire and Co., Ltd.)

conduct further development work, or to abandon the project altogether. Many companies maintain a laboratory section responsible solely for monitoring and conducting test activities.

DESIGN: SELECTION, SPECIFICATION

When we speak of engineering design, we must distinguish between the process of design and the function of design. Sometimes described as a morphology, or structured set of problem-solving activities, the process of design begins with the recognition of a need and ends with an engineering solution that will result in some new or improved product, a new method, or a one-of-a-kind device or system. The process of design may include the specific activities involved in research, development, design, and the other engineering functions yet to be discussed.

In contrast, the engineering design function is concerned with the transition between the development concepts, or less-than-optimum development models, and the mass production, or one-of-a-kind construction, of the product. The function of design, then, becomes a smaller set of activities, the purpose of which is to render a final design sufficiently detailed for mass production or construction. It is the step between development and final production or construction, one in which design engineers will closely interact with engineers from the other work functions, particularly development, production, and construction. As one would expect, management is intensely interested in this activity because a favorable design may require management to commit significant company resources to the continuation of the project or product development.

Design Description

The function of the designer is to provide report and drawing details complete enough to enable production or construction engineers, despite unfamiliarity with the design at hand, to economically produce or construct the item or system specified. This degree of preciseness requires the design engineer to be knowledgeable about materials, production and fabrication procedures, and construction and assembly methods; to be familiar with common engineering devices and components; and to understand the practices governing the graphical communication invariably required in the design detailing.

With this knowledge, the designer will then spend significant detailing time selecting and specifying appropriate materials, components, parts, sizes, tolerances, and so on, and any required processing, assembly, operational, and maintenance procedures. (See Figures 2.6 and 2.7.)

In any given field of engineering, the designer will have a wide range of practical experience and the ability to satisfy the many constraints that are always imposed on the final design. For example, the mechanical designer responsible for designing a new automobile body will select the materials and specify the sizes, shapes, assembly, and production methods, working within restrictions determined by the new styles, weight limitations, strength requirements, corrosion resistance, durability, ease of assembly, and economics of production. Suppose this new body is to be guaranteed against significant corrosion for three years. The designer is then faced with several economic choices: the use of a cheaper thin sheet steel subject to corrosion, which must be protected by the added expense of chemical coating; the use of an expensive noncorrosive stainless steel; the use of nonmetals such as fiberglass or high-strength plastics, which are also noncorrosive; or, finally, some combination of these. The designer faced with these choices must consider production and labor costs associated with the choices ultimately made. Labor costs associated with the production of a fiberglass body may contrast significantly with those for a steel body, which can be shaped into some particular configuration by stamping or bending. Then again, the curvature and overall shape specified for the body might exceed the expectations of the stamping operation, with fiberglass the only workable choice.

An architectural designer might be faced with determining the amount of insulation required for a new building. More insulation will reduce the initial expenditures and operating costs for the environmental control system but increase the initial expenditures for insulation. And suppose the environmental control designer also working on this building is asked to design a solar heating system. This will necessitate coordination with the architectural

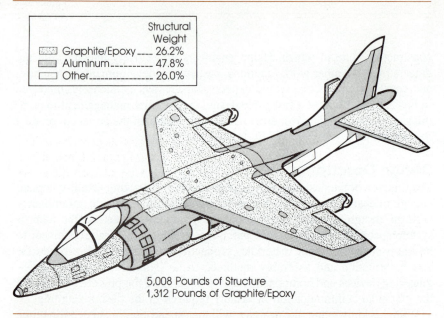

	Structural Weight
Graphite/Epoxy	26.2%
Aluminum	47.8%
Other	26.0%

5,008 Pounds of Structure
1,312 Pounds of Graphite/Epoxy

FIGURE 2.6

The need for greater thrust-to-weight ratios is causing engineering designers to rely more on composite materials. In this aircraft, composite materials account for 26 percent of the structural weight. (Reproduced with permission. © 1980, Society of Automotive Engineers)

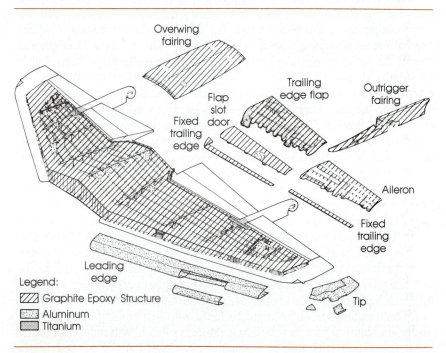

Legend:

Graphite Epoxy Structure

Aluminum

Titanium

FIGURE 2.7

The designer is faced with assembling the composite materials, some of which are not always compatible with one another. To do this requires detailed knowledge of assembly techniques, material properties, and the effects of hostile environments on the material behavior. (Reproduced with permission. © 1980, Society of Automotive Engineers) or civil engineers as the designer struggles with the basic decision of using an air or water heat-transfer medium. Since water has a higher heat capacity than air, the overall solar collector area could be reduced from that required for air, but the overall complexity and cost of the system will increase in order to protect against water leakage, mineral deposition in the piping system, and overnight freezing. Other considerations for the designer include the weight trade-offs between the different types of solar collectors required, depending on whether the medium is air or water. Ultimately these trade-offs affect the structural and aesthetic design required to accommodate either type collector.

Characteristics of Good Designers

A good designer must be able to make decisions when there is more than one way to accomplish the design purpose. A designer who must achieve some type of motion or transmit some force may be able to accomplish this by several means—electrically, hydraulically, pneumatically, or mechanically. Other designers may face different choices in making a part—whether to machine it, cast it, stamp it, or grind it. Still others may be faced with separating out a chemical or mineral element either by chemical reaction, flotation, electrolysis, or centrifugal action. Decisions of this kind obviously involve complex evaluation and sound judgment requiring years of experience. Fortunately, modern computer technology has provided improved computational aids for designers. Improvements in computational speed with corresponding reduction in system costs and system size have resulted in the technological development called computer-aided design or simply CAD. In many cases, it has allowed designers or supervised technologists to integrate experience, computer design programs (software), and terminal graphical displays in an interactive decision-making process. As a result, the designer can dynamically and visually select the best designs.

Good designers place the design in proper perspective and strive for simplicity. In view of the myriad of details accompanying any design, it is easy to lose direction and purpose. A designer who gets lost in the utilitarian aspects of a design and forgets the aesthetics of something made for public consumption may make decisions that prove disastrous. Proper perspective may be achieved by applying balanced design.

Theoretically, a system should last its intended life and provide satisfactory service—and no more. In reality, designers cannot accomplish this goal. Consider the American automobile, generally designed to provide good service for about 100 000 miles. It is unrealistic to design every part to last the intended life. Instead, it is possible to design the power train (engine, transmission, drive shaft, rear end, bearings, chassis, interior suspension system, steering mechanism, and basic electrical system) to meet the life expectancy. But designing parts such as automobile tires, V-belts, brake linings, spark plugs, batteries, or headlights to last 100 000 miles makes automobiles too expensive to buy or sell. Instead, automobiles are designed so that periodic replacement of these parts is necessary but at reasonable cost to the consumer.

In some instances a designer will include a system component or part that will exceed the expected design life for the whole system. Although this may appear to violate balanced design, sometimes it proves less expensive to specify an overdesigned "off-the-shelf" component or part rather than build a new one. The implication is that designers must continue to stay abreast of new technological developments in all engineering disciplines, particularly when it comes to parts, components, equipment, and the like.

QUALIFICATIONS   Engineering students interested in design should select a broad undergraduate education rather than a narrow one and should consider a master's degree in engineering. Note that only 3 percent of engineers with doctorate degrees work in design—insignificant compared with the 15 percent of those with master's degrees and 20 percent of those with bachelor's degrees. Because the designer becomes a focal point for measure of design progress, it is essential that designers increase their communication skills and abilities to work cooperatively and objectively with other engineers, managers, parts and equipment vendors, draftspeople, technicians, and technologists.

Production-Consumption

Whereas research, development, and design are primarily engineering responsibilities, the decision to produce or construct (often involving enormous economic commitments) is basically management's responsibility; thus, management shares these last phases in the life cycle of a project with engineering. Now we enter a final stage called **production-consumption**— quite distinct from research, development, and design, which constitute the initial stage called primary design or engineering design.

Management must now expand its boundaries of administration to include not only the engineers who work in production, construction, operations, maintenance, and sales but the many other personnel who support the planning for and carrying out of the distribution, consumption, and retirement of the design end product.

Depending on the type of product, the number of support personnel may far exceed the number of engineers involved in this last stage. Products that are marketed in large quantities—such as televisions, automobiles, and appliances—require large numbers of assembly-line workers, equipment operators, and other blue-collar laborers. In addition, business, legal marketing, and sales experts are engaged in monitoring and controlling the flow of company materials and financial resources while selling the product to the consumer. If these last activities are not administered carefully, even a well-designed product can flop in a competitive marketplace.

In the production-consumption stage, engineers are less concerned with the process of design than with the design of process. Once the design is more or less identified, questions such as the following are asked: What is the most economical way to produce or construct the design? What production or construction processes should be used, keeping in mind material and personnel limitations? What operational and maintenance procedures will produce the best production processes and prevent breakdowns?

QUALIFICATIONS   Engineering must play a decisive role in the production-consumption stage because production, construction, operations, maintenance, and sales are primarily engineering functions. In fact, 22 percent of all engineers work here, of which most have bachelor's degrees, some have master's degrees, and only a few have doctorates.

PRODUCTION AND CONSTRUCTION: PROCESSES, ASSEMBLY, STRUCTURES, SYSTEMS

Production and construction are similar engineering activities, the difference being in the type and quantity of end product. When we compare the production of automobiles with the construction of a highway or bridge, the production of televisions with the construction of a television station, or the production of electric appliances with the construction of a coal-fired power-generating plant, we delineate these differences. We see that **production** activities usually involve smaller items that are produced in quantity in plants or other specified, fixed locations. In contrast, **construction** involves the erection or assembly of structures, the integration of various component systems, and the solution of a set of unique problems arising from the varying conditions at each construction site.

In spite of these differences, there are many similarities between production and construction. Each is concerned with converting raw materials to a product, device, or structure at a price competitive with others on the open market, and at the same time meeting imposed profit, quality, and reliability constraints. Each involves a variety of specialists who contribute knowledge of materials, assembly methods, equipment, reliability and quality assurance, manufacturing methods, cost estimating and marketing, contracts, negotiations, and project scheduling. In addition, both can require enormous resource commitments and financial risks; thus, they are directed by management, which in turn delegates authority to production or construction engineers. Finally, both production and construction engineers influence the final design specifications by coordinating, according to time-tested guidelines, the production or construction planning with engineering designers (see Box 2.4).

Production

When a company decides to market a design, production engineering moves into high gear. Using the detailed design specifications, the production engineer through **process definition** integrates materials, people, production processes, and facilities into a system capable of delivering, on schedule, a finished product.

Box 2.4

Guidelines for Production, Construction Planning

The details of how a part or structure is designed generate its cost.
- A design that misses its cost target is usually a bad design.
- The prototype should be simplified to create a simple product or
- structure.
True reliability comes from design simplicity.
- Quality means strict technical adequacy—not frills.
- The design of key parts or systems includes the production tooling
- and equipment selection.
The design should promote assembly efficiency.
- An engineering design is not finished until it has been evaluated in
- its finished form.

After process definition, production engineers and technologists become concerned with facilities. Does this product require an entirely new facility? Is a new facility needed to replace an outdated one, or can the old one be renovated to provide for increased volume, redesigned goods and services, or new processes and equipment? If a new facility is required, then the production engineers must locate the facility, considering such factors as the regional demand for the product; transportation availability; utilities; local taxes; labor availability and costs, site and construction costs, tooling costs; local attitudes, schools, churches; and growth opportunities.

Next the production engineer or production technologist will be concerned with the layout of the facility, keeping in mind the objectives of minimizing materials-handling costs, reducing bottlenecks, reducing hazards, utilizing labor and space effectively, improving personnel morale, and providing for flexibility and ease of supervision. Depending on the volume, weight, size, and fragility of the product and on the type of equipment, the production engineer will have to choose a layout type. For smaller, less complicated products, the production engineer may select a flow or assembly-line layout to assemble the product step by step as it flows from one operation to the next. When the items are big or heavy, such as airplanes or ships, layout by fixed position becomes necessary. In these cases, completion of each production step necessitates bringing a new set of specialists and equipment to the item. Finally, facilities planning must incorporate room for materials, inventories, tools and equipment, first-aid offices, shipping and receiving areas, washroom and locker facilities, refreshment and cafeteria areas, and offices for personnel involved in maintenance, inspection, quality control, security, safety, and supervision.

QUALIFICATIONS Production engineers and technologists must meet a wide range of qualifications, from essential knowledge of materials, processes, equipment, and systems to a basic understanding of statistics, engineering economy, and human factors. They must be able to communicate with labor personnel, management, other engineers and technologists, and other company specialists in what can frequently be a high-pressure environment. They must be able to expedite repairs or other problems, such as labor disputes, which might hinder the flow of production items.

Engineering and technology students interested in production work should pursue studies in materials and manufacturing processes, learning about properties such as tensile, shear, and impact strengths; hardness; electrical and thermal conductivity; dielectric properties (whether the material is an insulator or a conductor); magnetic permeability; fatigue resistance; and corrosion resistance. They should learn how to enhance or change these properties and they should learn how materials respond to chemical changes, thermal fluctuations, radiation flux, static and dynamic stresses, and electromagnetic fields.

Basic production processes should be studied by comparing their production rates, economics of operation, accuracy, suitability for mass production, and effect on material properties. Certainly, students will want to understand the differences between liquid forming by casting or solid-state forming by forging, extrusion, rolling, and stamping. They should become familiar with the equipment and technology to remove materials, whether by machining, grinding, or drilling, and in some of the more precise methods using electric discharge, electrochemicals, ultrasonics, or laser-machining equipment.

Courses in systems control, mathematics, and computers are desirable because of the highly automated and controlled production environment. To prepare for reliability and quality control, courses in statistics and probability should be taken along with laboratory courses demonstrating modern inspection techniques. Courses in social sciences, English, communications, and psychology are also important.

Construction

It is estimated that the construction industry alone accounts for approximately 12 percent of the gross national product (GNP) and about 15 percent of the total U.S. employment. Numerous opportunities are provided construction engineers as they engage in the actual work of building structures. Most will be employed in the "heavy" construction of large buildings, power plants, refineries, highways, airports, bridges, and walkways. Others will work in the construction of residential buildings, small nonresidential buildings, and public utilities.

The construction process is initiated when some agency or sponsor puts out a request for bid on a structure or system it desires built to certain specifications. It is then up to a general contractor or a team of contractors to bid in competition with other contractors for the job.

The work of construction engineers and technologists begins once the decision to estimate and bid on a job has been made. Their knowledge of construction procedures, machinery, and structural materials and their ability

Precisely milling nearly 50 tiles of silica insulation simultaneously, this machinist demonstrates the advantages of numerically controlled milling machines. The tiles will be fitted together like a giant jigsaw puzzle to form a heat shield to protect passengers aboard NASA's space shuttle. To ensure aerodynamic smoothness in flight, the huge milling machines are programmed to match the bottom of each tile perfectly to the curvature of the shuttle surface at the exact point it is to be attached. (Courtesy of Lockheed Missiles and Space Company, Inc.)

to assess site characteristics and their effect on costs are essential to realistic costs estimation.

It is the judgment and experience of the construction engineers that answer many questions about structural components and where they should be made; about supporting facilities and where they should be located; about special problems caused by the site location, such as unusual topography, soil conditions, or too much or too little ground water. For example, jobs with extensive concrete work need readily available supplies of sand, gravel, and water if the work is to be done on or near the site. The construction engineer determines whether a concrete batch plant will be erected nearby or if the concrete components can be fabricated off-site and transported in for assembly. Other questions may arise regarding special transportation systems: Must they be

These engineers are checking an assembly detail from the construction plans for a new chemical refinery. (Courtesy of Exxon Corporation)

built to transport components and materials to and from the construction site? Other questions may relate to power availability or whether generating equipment is to be transported in. Do the local soil conditions permit the movement of heavy construction equipment, or will special roads have to be built?

Because of these uncertainties, the construction engineer or technologist expects every construction job to provide new and unique experiences. Almost every job integrates a new team of architects, subcontractors, engineers, and sometimes technologists, as well as different financial, legal, insurance, and government representatives. The local topography, weather, transportation facilities, material accessibility, utilities, and labor conditions will vary tremendously from site to site. Obviously the construction conditions associated with the Alaskan pipeline were considerably different from those associated with drilling platforms in the North Sea.

FIELDWORK   Once the project is under contract, the construction engineer or technologist will engage in either fieldwork or office work and sometimes both, depending on the size of the construction job. Fieldwork usually begins with a site survey and layout of the project. Then it moves to the construction of access roads and auxiliary buildings, if needed, plus other excavation work. Then the structure erection begins with the aid of modern construction methods and equipment. There can be tremendous variation in the equipment and assembly methods, depending on the type of construction. In the mechanical construction of systems such as refineries, the construction places more emphasis on the assembly of prefabricated units rather than building in place. Piping, ductwork, and control systems play even more important roles in mechanical construction than in heavy construction, where the object is to support loads, transport large objects, or house people.

OFFICE WORK Drafting, accounting, calculating, plan and specifications interpretation, contract clarification, and progress reporting absorb the time of the construction engineer or technologist engaged in office work. Note that both field engineers and office engineers or technologists will write program reports and act as troubleshooters should labor disputes, subcontractor difficulties, or right-of-way entanglements arise.

QUALIFICATIONS The qualifications of construction engineers and technologists are similar to those of production engineers and technologists, especially in the areas of people interaction and personal communication. Students interested in either engineering function need, as a minimum, a bachelor's degree. Students will be required to take courses in materials, economics, social studies, communication, fabrication methods, and mathematics. For those interested in construction, training in construction theory, methods, and equipment would be worthwhile, if available.

OPERATIONS AND MAINTENANCE: PERFORMANCE, SERVICE, REPAIR

Operations and maintenance, its natural partner, are basically concerned with the supervision, control, and upkeep of products and production systems in both manufacturing and nonmanufacturing activities. When products or systems become sufficiently complex, the technical knowledge of operations engineers and the skills of operations technologists are needed. For example, in nuclear plants producing electric power, operations engineers supervise the complex plant operations involving the delicate energy interchange that converts radiation flux to steam energy, steam energy to turbine mechanical energy, and mechanical energy to generated power. Technicians trained and supervised by operations engineers or technologists will probably do the actual monitoring and adjusting of the complex control systems, but it will be operations engineers who analyze and interpret the operational data and then determine the necessary adjustments, always striving for increased efficiency. Operations engineers or technologists also supervise maintenance specialists, schedule inspection and maintenance procedures, and supervise the installation of new equipment.

In manufacturing plants the operations engineer's role may expand somewhat from that in a utility plant as the variables become less defined. Oper-

An engineer and a technician make a last-minute check on a mass storage subsystem at a manufacturing plant. Quality control requires constant attention to detail. (Courtesy of IBM)

ations engineers or technologists are usually involved with production engineers in production planning, particularly in the layout of facilities. Following initiation of production operations, the operations or "plant" engineer or technologist responds to work orders based on demand forecasting. Then he or she requests materials from inventory control, schedules personnel by dispatching work orders, schedules production, and finally measures the effectiveness of the production operation by analyzing the final product, the labor costs per item, and any schedule slippage and by monitoring the raw material and product inventories along with any scrap material.

Other Responsibilities

Operations engineers and technologists also assume responsibility for maintaining plant safety; planning for expansion; organizing resources, including the scheduling of people, equipment, and time; directing, training, and motivating people; and laying out and assembling new processes and production equipment. Operations engineers will ensure product quality by directing quality control specialists, often technologists, as they assess the effectiveness of the production processes and equipment by determining the quality of the end product. This activity of **quality assurance** seeks improved production reliability while minimizing the costs to produce the item. Quality assurance also evaluates whether the product is convenient, is safe to use, and conforms to the manufacturing specifications. It will be up to the operations engineers to use the quality assurance data to improve production procedures, order new equipment, or realign the facilities and improve efficiencies in production.

The sophisticated electronics for this Space Shuttle crew trainer require continual maintenance. The engineer here must sort out a flood of facts as he checks and calibrates the instrumentation. (Courtesy of IBM)

Maintenance

Operations engineers and technologists also organize and supervise the maintenance of buildings, equipment, grounds, and utilities. In technologically complex systems, proper maintenance procedures are absolutely critical if a company is to reach its design and profit goals. A jet airliner grounded for unexpected repairs costs the airline thousands of dollars in lost revenues. An offshore platform unable to drill because of diesel engine failure loses the drilling company about $40 000 to $50 000 a day. Conveyor failure in an assembly-line production plant may slow down the entire production system, causing labor to stand around, wasting energy and causing schedule delays.

For these reasons, operations engineers set maintenance objectives and organize to meet them to minimize the loss of production time because of malfunctioning equipment or poor maintenance. They plan efficient use of maintenance personnel and equipment to protect the company's investment in its buildings, plants, and equipment. They select and install equipment and schedule preventive maintenance based on some predetermined and acceptable reliability. They see that maintenance personnel keep records of periodic inspection and testing, lubrication, painting, cleaning, adjusting, and other servicing requirements. Operations engineers will make decisions regarding contract versus in-house maintenance by considering department size requirements, overhead, equipment costs, and plans for facility expansion.

QUALIFICATIONS Operations and maintenance engineers and technologists are less concerned about abstract concepts in engineering than with the fundamentals governing production and operational systems. Most will have a bachelor's degree and will need a broad background obtained from studying subjects in the other engineering branches such as chemical, civil, electrical, industrial, materials, and mechanical engineering.

SALES: MARKETING, APPLICATIONS, SERVICE

Have you ever shopped for an item only to discover that the enthusiastic salesperson could not explain any of the technical details that you needed for your personal assessment of the product? This can be a most frustrating situation, one that engineering has tried to eliminate through the function called **sales engineering.** The increasing complexity found in engineering products and equipment combined with the diversity of industrial operations has caused industry to dedicate engineering resources to the marketing activity. Consequently, almost 5 percent of all engineers work in sales engineering.

Beyond the obvious sales function, sales engineers usually fulfill company objectives of **enlightenment, applications, service,** and **company representation.**

Enlightenment

One of the purposes of sales engineers is to introduce and demonstrate company methods, products, equipment, and capabilities to the marketplace, that is, to the consumers of the industrial technology. For example, a company develops and is ready to market a new microsonic sensing system that is capable of detecting nearly any type of object passing through its beam path. The sales engineer will make appointments with those industrial representatives or consumers who may be interested in this type of product—in this case, companies that utilize modern control systems in assembly-line production. At the meeting, probably with production or operations engineers, the sales engineer points out the favorable characteristics of the sensing system, noting that the system probes utilize a beam with invisible, inaudible, ultrahigh-frequency waves that allow the detection of objects as small as 0.5 inch moving through the beam at a rate of up to 2 000 objects per minute. The sales engineer also notes that the sensing beam does not involve physical contact and that the beam's narrowness makes the system immune to disturbance by light, dust, water vapor, and other particles that might pass through the beam. Finally, the sales engineer may note that the system detects either metallic or nonmetallic objects and has a 15-volt circuitry and output signal that makes it directly compatible with many logic and control systems. It is hoped that this information will lead to the sale of the product.

Applications

It is up to the sales engineer to become familiar with those engineering representatives who make decisions about new products and equipment. By continuing interaction with these representatives, the sales engineer strives for familiarity with each company's production operations. Opportunities may then arise when the company is considering new equipment for replacement or expansion or when the company has operational difficulties that necessitate redesign of operations. Sales engineers may then actually engage in system design by proposing methods and equipment to the company engineers to accommodate the necessary operational changes. Sales engineers will need a thorough knowledge of the capabilities and limitations of their own products and also an understanding of the capabilities of many other products from other companies. The engineer will then propose a system, in competition with those proposed by other sales-engineering representatives, to accomplish the necessary objectives, often including parts and equipment from other manufacturers.

It is not unusual for needs to arise for which no process or product exists, or for which minor design changes must be made to make the process or product suitable to an industrial firm's immediate requirements. Thus, the sales engineer is continually feeding back design information to his or her company that may lead to a new development project or to temporary design modifications.

Service

Marketed equipment and systems arc often so complex or sensitive that the industrial consumer chooses not to maintain them; thus, the sales engineer often establishes a maintenance contract. With this added responsibility of maintenance, the sales engineer schedules the service department, usually composed of technicians and other specialists, for preventive maintenance and supervises any repair work that occurs, always looking for design defects. The service department will also assume the field testing of newly installed equipment.

Company Representative

Sales engineers reflect a company's reputation, including its technical prowess, its reliability, and its overall dedication to excellence. The sales engineers not only must be technically competent in design, production, operations, and maintenance, but also must be dependable, friendly, courteous, tactful, able to listen, and able to inspire confidence in their company's products. Therefore, sales engineers need to be good communicators and able not only to listen but also to interpret and express ideas properly. This "sowing of seeds" as company representatives requires sales engineers to travel considerably.

QUALIFICATIONS Because sales engineers work on all kinds of problems, their education should be broad, similar to that required for production and operations engineers and technologists, culminating in a bachelor's degree. As details about company products will not be taught in college, sales engineers will need apprenticeship time in either manufacturing or service departments to gain the knowledge and judgment skills necessary for sales engineering.

It is the basic function of product support to give customers the utmost assistance in the operation and maintenance of such equipment as the aircraft engine you see here. These engineers maintain close contacts with airline and military engineering and maintenance personnel; thus, they play a key role in the continuing refinement of engine designs. Here engineers study the impact an engineering change will have on product support. (Courtesy of Pratt and Whitney Aircraft Group)

MANAGEMENT: PEOPLE, PROJECTS, PRODUCTS

Because the engineering design process, by its very nature, requires management, all engineers and, to a certain extent, technologists will play the role of manager during their careers, even though it may not be a permanent role or reflect an engineer's major work function. Management may be vested in the project engineer who is responsible for the foundation work of a high-rise building, or it may be vested in the engineer who directs a research project requiring the management of test equipment, technical assistants, computer

Engineering management is concerned with project planning and schedule deadlines. Here a group of project managers meets with top management to report on project progress. (Courtesy of Pratt and Whitney Aircraft Group)

usage, budget disbursements, and overall research progress. The engineering technologists will often manage technicians, craftspeople, and other nontechnical personnel in applications projects.

In some cases engineers will move into permanent supervisory positions or even to higher-level company management, thereby assuming the overall responsibility for project or product success. Today the trend is to move more engineers into the higher management of industrial firms. In fact, 30 to 50 percent of industrial managers will have an engineering background. This is in sharp contrast with the turn of the century, when less than 10 percent of the top management had engineering backgrounds.

People Management

Engineering managers must have the ability to supervise people (the personnel) and work with higher management. This may require the manager to establish staffing plans, write job descriptions, and establish promotion and wage and salary guidelines. Managers may have to manage taxes and employee records and to establish and regulate working hours and fringe benefits such as vacations, insurance, and retirement. Some will be concerned with establishing programs and training for engineering personnel development to promote technical effectiveness and increase personnel morale. Since engineering managers will work with many different people, they must learn how to communicate well and how to be sensitive to the needs of those they work with.

Another important characteristic that engineering managers must possess is the ability to organize and plan. Without this crucial ability, engineering managers would often be placed in jeopardy. Managers, therefore, are usually familiar with scheduling techniques such as PERT (program evaluation and review technique) and line of balances; these enable managers to build contingency steps into the overall organizational structure for protection against unexpected delays. One cannot minimize the importance of experience in this area of project management.

Engineers who become full-time supervisors or managers are usually

- Technically proficient and experienced
- Capable of making sound engineering judgments and decisions
- Able to motivate and work with people of varying backgrounds
- Proficient in costs management and budget control
- Able to understand basic contract law and to negotiate contracts and mediate disputes
- Able to set goals and objectives and organize and plan to meet them
- Responsive to directions from supervisors and top management
- Able to operate efficiently in an environment of risk and uncertainty

As would be expected from this list of characteristics, engineering managers, particularly those at the higher levels of management, are usually paid more than engineers involved in the other work functions mentioned earlier. Note that engineers require job experience and established technical competency before they are given opportunities to manage. Young engineers should not expect to move into top-level management immediately.

Project and Product Management

Engineering almost always results in something tangible. It may be a one-of-a-kind project item such as a communications satellite, a new dam, or a new irrigation system. Then again, it may be a product that is manufactured in mass for commercial distribution. But regardless of the end result, it cannot be successfully constructed or produced without the careful management of a series of intermediate steps that govern movement through the overall design process.

As an example, consider the major project of building a dam. Here the management of field operations and the supporting projects is crucial to overall success. Among the many initial projects that may be necessary are the site survey and the establishment of field site support, including the facilities for housing, meals, school, equipment storage, and medical operations. Managers would direct the efforts of others to establish communications, construct roads, and distribute power to the various facilities; or they may be involved in task planning requiring statements of work, permit clearances, and procedures for material and equipment accounting, handling, and inspection. As the field operations progress, managers may then become more involved in contract control, scheduling of action meetings, quality control, auditing, and reporting.

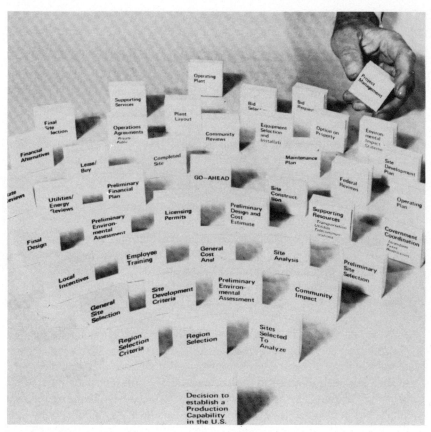

Decisions, decisions, decisions! Engineering managers live with them every day. (Courtesy of the Boeing Company)

Societal and Fiscal Accountability

Since success and profit are often considered one and the same, the engineer or technologist must build a device or manufacture a product that meets the requirements established for it at the projected costs. The engineering manager is often concerned with the flow of development and production funds while ensuring that the end product performs satisfactorily. In fact, engineering managers today assume a significant responsibility when it comes to the consuming public. If company products perform unsatisfactorily, damage the ecology, or lead to injury, death, or sickness, then the principal management becomes accountable for the company as its official representatives.

QUALIFICATIONS The level of responsibility and enlightenment will increase as the engineer moves into higher management. All engineering managers must understand the basic budgeting and accounting procedures used in everyday financial operations such as accounts receivable, accounts payable, billing, and payroll disbursement. But those who become engaged in the management of overall product development will have to understand basic cost-structuring principles that integrate the various costs of planning, research, design, production, marketing, and operations to determine a unit cost for the product. Managers here may have to understand tax administration and the stock and dividend structure, know when to make capital investments, and be familiar with the methods of financing. Most engineering managers will have at least a bachelor's degree in engineering and many will have advanced degrees in engineering or in business administration.

LOOKING BACK

One thing should be evident from this chapter and the one preceding it: engineering is diverse. There are well over twenty major branches or fields of engineering, twelve of which were discussed in some depth in the previous chapter. (Others such as biomedical, energy/power, environmental, geological, and systems engineering are no less important but are just not employing large numbers yet.) In addition, every field of engineering has several specialty areas. For example, chemical engineering has at least seven, and mechanical engineering has eleven or more.

This diversity is increased even further by the different functions that engineers can perform, including management, research, development, design, testing, production, construction, operations, maintenance, sales, consulting, and teaching. Most of these functions represent important activities in the process of design, which occurs in basically two stages, primary design and production-consumption. Most engineers will work in management, research, development, and design, although there are many opportunities for those interested in the less abstract engineering activities of production, construction, operations, and sales.

Combining the number of engineering fields, the more specialized application areas of the engineering technologists as described in Chapter 1, and the number of specialty areas with the number of engineering functions, we can count thousands of different engineering "niches" providing unique opportunities and experiences for prospective engineers and technologists. Neither are forced to remain in these niches and will probably move from one to another during their careers.

Engineers and technologists must work in a dynamic environment, requiring that they have a broad education, along with some training and experience. They will work with ideas, beginning either as abstract concepts or principles or as the result of synthesis or innovation; work with machines, equipment, and structures; work with methods and processes; and work with people, including other engineers and technologists, management, technicians, employees, customers, and representatives from both the public and government sectors. Communication is absolutely necessary if the engineer or technologist is to take ideas and make them into reality. Finally, engineers and technologists work with money (project budgets, finances, capital investments, utilities, material costs), realizing that it is the economics that often dictates the engineering decision.

PROBLEMS

2.1 Interview an engineer in industry whose primary responsibility is management. Discuss the characteristics that engineers must have before they can assume management roles.

2.2 Interview a civil engineer or technologist in construction who has recently completed a significant construction project. Ask this individual to relate the more difficult aspects of the project, including not only the engineering problems but also the people problems.

2.3 Investigate one of the more recent research breakthroughs. Determine the more important intermediate steps that led to this important breakthrough.

2.4 Engineering researchers are at the top of the "abstract" scale in applying abstract principles to their work, whereas management is more likely to be at the bottom of this scale. With management at the bottom and research at the top, place all the engineering functions in descending order in their application of abstract principles.

2.5 Interview one of your engineering professors and ask him or her to relate why he or she chose teaching as a career.

2.6 List the educational prerequisites required for each engineering work function.

2.7 List the aptitudes that are desirable for engineers in each work function.

2.8 Discuss the differences and similarities between production and construction.

2.9 What are the differences between development and design? Is testing a more important activity for one of these work functions? Explain.

2.10 Generate several new design ideas that you believe are worthwhile projects. Take the most appealing one and conduct enough library research to ensure that a design solution is not already in existence. Report on your findings.

2.11 Compare the present advantages and disadvantages of electric automobiles over automobiles powered by internal combustion engines.

2.12 Different kinds of vehicles are used in different transportation situations. Choose a type of vehicle for each of the following transportation situations and explain your selection in each case:

a. the delivery of a householdful of furniture across town
b. a hunting trip in a wooded mountainous area
c. a 1000-mile camping trip to the western coast of the United States

2.13 List five new engineering designs that are needed to prevent burglaries.

2.14 Select a common waste product and suggest ways it can be reclaimed or reapplied for useful purposes.

2.15 List at least five desirable properties (for example, strength, low cost, and light weight) that design solutions must exhibit in each of the following design situations:

a. a four-person golf cart
b. a trailer that can be towed by bicycles
c. a backpack used by hikers in the Arctic regions
d. a portable solar cooker

2.16 List the design requirements and restrictions for a bomb shelter constructed for an average-size family. Include the requirements and restrictions relative to size, food, water, stay duration, ventilation, waste, lighting, maintenance, and location. Develop a design that meets your proposed design criteria. Have someone else (your instructor, a friend, or an engineer) evaluate your preliminary solution.

2.17 It is desired to build a tent that is structurally sound in shifting high winds of up to 80 kilometers per hour. Without first determining all the criteria for the design, propose several design solutions and make sketches of each. Now develop the design criteria (normally done *before* design solutions are proposed). Have friends assist you in this development. Compare each of the proposed solutions with the design criteria. Based on the results of this comparison, discuss the advantages and disadvantages of developing design criteria *before* proposing design solutions.

2.18 Figure 2.8 presents a humorous approach to an improved wheelchair. But is it possible that some of the design features shown are not so farfetched as they may seem? It is not unusual for wild ideas to stimulate more practical ones—this is an important principle of brainstorming. After examining each of the design features in the figure, list all the ideas for an improved wheelchair that are stimulated by your examination. Discuss the design implications of each of these ideas.

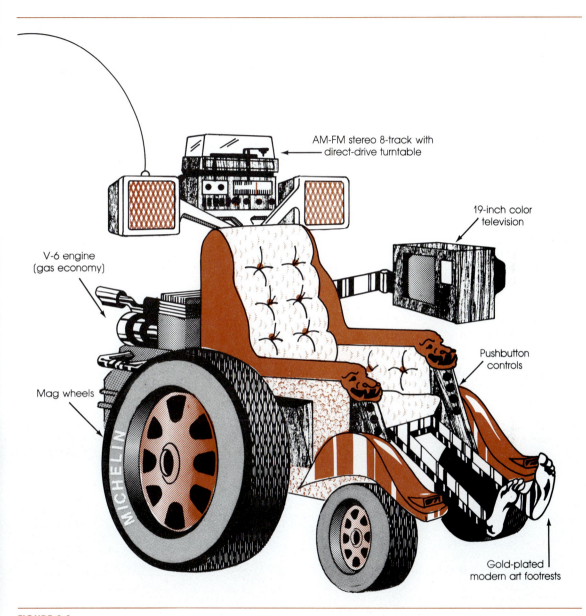

AM-FM stereo 8-track with
direct-drive turntable

19-inch color
television

V-6 engine
(gas economy)

Pushbutton
controls

Mag wheels

MICHELIN

Gold-plated
modern art footrests

FIGURE 2.8

An improved wheelchair?

2.19 An engineering design requires a tube of 1 meter in length having an outside diameter of 3 centimeters. The design requirements specify that copper, steel, and aluminum are acceptable materials. Which is the least costly material, assuming that the design analysis determined the minimum inside diameter of the 1-meter tube to be 2.4 centimeters for copper, 2.6 centimeters for aluminum, and 2.8 centimeters for steel?

The price per kilogram of metals will vary considerably from one material to the next.

2.20 Explain how the computer might be applied in the design of each of the following:

a. an improved football helmet
b. a slurry pipeline (a pipeline that transports small coal particles by suspending them in flowing water)

c. a new automobile suspension system
d. a highway that must pass through a mountainous region

2.21 List all the human-engineering problems that you can identify for the design of a baby stroller.

2.22 List all the human-engineering problems that have frustrated you in the use and maintenance of your automobile.

2.23 Make a list of five designs now marketed that you consider are not in the best interests of the con-sumer and express reasons for your convictions. Iden-tify the needs for which they were designed and propose modifications (if possible) that would resolve the more serious design deficiencies of each.

2.24 List five designs that are not aesthetically pleasing to the eye and express reasons for these deficiencies as you perceive them. Ask others to ex-press their opinions of the aesthetics of each of these designs. Compare their perceptions with your own and discuss any differences.

PART TWO
Preparing for a Profession

CHAPTER 3
PURSUING A CAREER IN ENGINEERING OR TECHNOLOGY

CHAPTER 4
THE PROFESSION

3 Pursuing a Career in Engineering or Technology

CONTENTS

Is Engineering or Technology for Me?
The Education
Preparing for a Profession
Entering the Profession
The Professional Years

You are entering the most difficult phase of your educational life. And while it is true that engineering or technology education is demanding, it is also true that the rewards are great. Do not let the difficulties of education cause you to lose sight of the satisfaction that will come to you as you are born into the profession.

IS ENGINEERING OR TECHNOLOGY FOR ME?

Is engineering or technology for me? If it is, in what branch or field or discipline? Students usually try to answer these questions by their sophomore year of college. Otherwise, they gamble time and financial resources educating themselves for a profession that may not prove compatible with their interests and abilities. Unfortunately, because of the numerous and at times intangible factors governing an individual's career decision, this chapter (and this book) cannot answer all the questions a prospective engineering or technology student may ask. It is hoped, however, that it will answer some, or at least establish directions for further search.

Technical training prepares people to perform well in any endeavor; students themselves recognize this. Increasing numbers of premed and prelaw students now prepare for medicine and law by first pursuing undergraduate programs in engineering. In addition, more engineering students are pursuing the master of business administration degree (M.B.A.), either by modifying their engineering curriculum content while extending their undergraduate engineering program one to two years or by entering graduate schools of business. This is consistent with the fact, as discussed in Chapter 2, that more and more engineers are moving into higher management.

In the past it was easier to identify the characteristics that engineers or students in technology seemed to display as a group. Engineering had not diverged into so many areas, the environmental and legal ramifications of engineering decisions were not so pressing, and the state of the art had not expanded beyond the individual engineer's saturation point. It was not uncommon for engineers or technologists to spread their abilities over several engineering functions, and the move between engineering branches could be made with a minimum of retraining.

85

The Age of Specialization

Today there is an increasing tendency for engineers and technologists to move toward specialized areas; and there are many to move to. For example, one engineer might be an acknowledged laser expert engaged in the development of some new laser weapons guidance system for an advanced military aircraft. At the same time, other technical specialists will be coordinating the control systems development required for the complex integration of the laser guidance hardware with the many other aircraft systems. There will also be experts to supervise the prototype construction, others to test it, and still others to guide the selection of materials, production procedures, and hardware that is ultimately used in mass production. And these represent only a sprinkling of the experts required for a project of this magnitude.

The proliferation of engineering specialty areas, clearly the result of rapid technological expansion, makes defining the prerequisite characteristics for a technical career difficult. For example, students who are interested in the relatively new biomedical engineering branch may not have been interested in engineering as it existed forty years ago. Today, students who enjoy chemistry, biology, and anatomy courses may enjoy the engineering challenges of developing new devices and equipment to sustain and monitor biological functions.

The set of qualities and characteristics listed in Box 3.1 should not be considered necessary prerequisites to pursuing an engineering or technology career. In fact, many of them are prerequisites to careers in business, law, medicine, and mathematics. Nevertheless, any practicing engineer or technologist will exhibit many of these qualities.

Early Preparation

An individual who satisfies only one or two of the characteristics mentioned may not find fulfillment in a technical career. However, such an individual could continue to learn more about the profession by talking to practicing engineers, reading engineering books, taking engineering field trips, and participating, if possible, in the summer career programs frequently offered by local colleges, many of which are sponsored by colleges of engineering and technology.

Prospective engineering and technology students should take all the high school mathematics and science courses available to them, including algebra, trigonometry, geometry, calculus, biology, chemistry, and physics. Other courses in English, history, social science, typing, and speech would be helpful, along with courses in mechanical drawing and shop. With these kinds of courses, the transition to college work will be much easier.

THE EDUCATION

Many students enter college with high expectations, having great confidence in their abilities. After all, they have excelled in high school, usually making

---

Box 3.1
Characteristics of Engineers and Technologists

- Has well-rounded high school career and performs well in mathematics and science courses.
- Has intellectual curiosity about how things work or how to make them work better.
- Has aptitude for working with physical things.
- Desires to serve humanity and shows interest in people.
- Has ability to communicate and work harmoniously with people.
- Is willing to work hard and to complete projects once they are initiated.
- Is able to assume and discharge responsibility.
- Is willing to accept and live by codes of ethical conduct.
- Enjoys learning and perceives the unknown as a challenge.

---

better grades than the average student. And although they may have been warned of the challenges ahead, most students fully expect to achieve at a level comparable to their performance in high school (and many will), even if it does require a "few additional study hours" a week. But it isn't long before the realities of this new educational situation become apparent.

Adjusting to the College Classroom

Students quickly discover that the classroom environment is significantly different from that of high school. They may discover that they are integrated with several hundred students in a general lecture class in which a professor delivers a formal presentation. After the lecture, they may next enter an informal laboratory or problems class supervised by graduate assistants. Within these different learning situations, some instructors may appear to cover course material too fast, while others may move too slow. Then some may appear rather aloof to student questions, while others are openly receptive.

Some instructors may take roll every day, while others do not. Some may give several exams during the term, others one or two; still others may choose to judge student capabilities through papers or reports, either written or oral. Instructors may seem to apply different performance standards or utilize different learning strategies, and they may or may not find the time to establish personal relationships with their students. So within this uncertain environment, students must choose, when choices are available, those classes and instructors that best serve their educational needs.

Educational facilities for engineering and technology students usually offer a challenging and well-equipped environment for experiential learning. This laboratory is dedicated to instruction in microprocessor technology. (Courtesy of Milwaukee School of Engineering and ASEE)

Exams and Grades

For some students, the next rude awakening comes in the first series of exams, when they realize that the classroom environment may be more competitive than in high school, that they are competing for grades against other students who have also excelled. This becomes even more discomforting when the instructor informs the class that grades are based on relative class performance curves or percentages. Other instructors may relate grades to predetermined levels of performance that demonstrate attainment of course objectives. In any event, after the first battery of exams, some students may be shocked by their unexpected low grades; engineering and technology grades usually average in the middle to high C's.

After the first few weeks, many engineering and technology students discover that those "few additional study hours" have now intruded into their usual recreation time. In fact, the majority are averaging two or more hours of outside study for each hour in class. For a fifteen- to eighteen-credit course load, this may amount to twenty-five to thirty-five hours of outside study a week. And depending on individual goals and personal preparedness, some students may spend considerably more time studying.

Putting It All in Perspective

It is at about this point that some students become discouraged. They may feel that the demands are unreasonable or that they are being forced to make decisions regarding personal organization and personal responsibility, decisions they did not expect to be confronted with, at least not this soon. Yet the road to an engineering or technology degree really isn't that steep if students will just step back to gain proper perspective.

Out of a 168-hour week (24 hours $\times$ 7 days), full-time students, on the average, will be in class and lab about 18 to 22 hours, which leaves 146 hours for other activities. Allowing 7 to 8 hours a night for sleep, 25 to 35 hours a week for study, and 4 hours each day for meals (and travel, if the student lives off campus), there are still over 30 hours a week for part-time jobs, student organizations, and other cultural, social, and recreational activities. Effort and organization of time are what a student really needs if he or she is to succeed.

Educational Goals

Entering engineering and technology students should determine what goals they expect to reach by the time they graduate. Is it the highest grade-point average possible? Or multiple degrees? Or a well-rounded learning and social experience? Goals such as these are not always compatible. For example, a student wanting to graduate with bachelor's degrees in both mathematics and engineering may have to take course overloads that ultimately lower his or her overall grade-point average. Yet, because of the additional degree, this student may be very attractive to an industrial firm specializing in the design and development of computers. Some students take challenging elective courses that promote broader individual development but may lower grades; others take easier electives in order to have more time for required technical courses. Many students participate in campus organizations, student government, and intercollegiate athletics.

Rather than dictate goals for engineering and technology students, the following sections present guidelines that should enable students to achieve academically and, even more important, to learn. These guidelines on personal organization, planning, and curriculum management should support the goals that engineering and technology students have. Graduating engineers and technologists should expect to have achieved the goals listed in Box 3.2.

Organization and Planning

Most students will have to change their living and study habits if they are to achieve the optimum learning experience. Attitudes toward learning, personal organization, and planning will evolve toward those the students will need in their profession. This is inevitable. But why shouldn't students speed this process up by organizing and planning right away?

Box 3.2

Goals of an Engineering/Technology Education

On graduation, the engineer or technologist will:

- Understand and be able to apply the fundamental laws of science and mathematics.
- Have a body of specialized technical knowledge and skills.
- Be able to solve problems effectively.
- Be able to communicate and work well with those within the engineering profession and with society in general.
- Be able to recognize and be concerned about the societal and political implications of technological achievements.
- Understand and abide by the codes governing the ethical conduct of engineers.

Organization and planning will prove most fruitful when applied to daily learning activities. Almost as important is the way engineering and technology students approach their curriculum requirements. Efficient curriculum management demands that students adopt an organized approach to satisfying their curriculum requirements, starting in their first term and continuing until graduation. This is especially true in engineering, as course requirements leave little room for manipulation. The curriculum, mostly technical, is arranged in building-block fashion with a foundation of "core" courses meant to be taken in a natural learning or prerequisite sequence. For example, students usually take calculus only after they have completed courses in algebra, trigonometry, and sometimes geometry. And students in civil and mechanical engineering cannot take engineering statics (the study of rigid bodies in equilibrium) before completing courses in elementary calculus and physics. Statics then becomes a required prerequisite for engineering dynamics (the study of bodies in motion under the action of forces). Dynamics is then used as a required prerequisite for other upper-level courses such as vibrations, a course often required in the mechanical engineering curriculum.

Freshmen curriculum requirements are nearly the same for all engineering and technology students. Courses in calculus, chemistry, physics, and English are required along with engineering courses covering subjects such as computer programming, mechanical drawing, introductory engineering analysis and design, and professional orientation. Because of this commonality, students can usually delay the decision about what to major in until their sophomore year.

Curriculum Management Guidelines

Engineering and technology curricula, formidable as they may appear, enable students to practice their profession following graduation with bachelor's degrees. This is in contrast to other professions, such as law, medicine, and dentistry, which require additional study and training in professional schools. Thus, the curriculum assumes a greater role in the development of engineering and technology students.

REVIEWING THE CURRICULUM   Beginning students should review the freshman curriculum listed in the institutional bulletin or catalog to determine whether they are sufficiently prepared for the required courses. Results from aptitude tests, prerequisite requirements, and advice from counselors can guide these decisions. When preparation deficiencies exist, students may have to take several developmental courses in basic mathematics and English before they pursue the normal curriculum. There are also mechanisms for better-prepared students to test out of some of the basic courses, usually by proficiency exams. Students who have taken calculus or mechanical drawing in high school may fit into this category.

While reviewing the curriculum courses, students should survey each course description to ensure that they have satisfied any prerequisites stated for them. If it is an elective course, then the course description is even more important. Identify courses that are prerequisites to those that might be taken the next term. It is crucial that engineering students get these out of the way first. Students usually have more freedom in the selection of humanities and social science courses, much less in math, science, and engineering courses. Obviously, these arguments apply equally well to the upper-level curriculum. Review the curriculum and course descriptions of every term. Students should note electives and choose those that satisfy educational goals.

COURSES ARE NOT ALWAYS OFFERED   Courses described in the college catalog may not be offered every term. This is particularly true of technical electives not always in demand by engineering and technology students. Students who need technical electives for specialized study should plan ahead by determining when those courses will be offered and then schedule their more flexible courses accordingly. Students who enter college at midyear (the spring term) usually have more schedule and prerequisite conflicts than those who begin in the fall term. They may be unable to schedule even required courses because course offerings are normally coordinated with a curriculum sequence meant to originate in the fall. Midyear students may be able to reduce scheduling problems by "phasing-in" with the intended curriculum by taking courses in summer school.

BALANCING THE TERM LOAD   Students normally take fifteen to eighteen credit hours of study a term. They will spend at least fifteen to eighteen hours a week in class and can expect an additional three to six hours in labs. If students carefully manage their curriculum, they can meet these demands and graduate as scheduled, usually in four to four and one-half years, sometimes five. To achieve this, students should try to balance their term load with a proper mix of required courses and electives. To take all technical required courses one term and all electives another term is not advisable, even if prerequisites are not violated, because technical required courses, on the average, demand more time than elective courses. Overloads greater than eighteen to nineteen credit hours are also discouraged because of the "rob Peter to pay Paul" principle. It is very important that engineering and technology students view each term as a new learning experience and that they get the most out of each class and lab, for these few years of learning and training are not easily replaced.

Guidelines to Improved Learning

Engineering and technology students, on the average, are more pressured than other students. From the beginning, regularly scheduled classes and laboratories, homework assignments, exams, lab reports, and other activities seem to compete for the available time. It is in these daily battles that students either become committed to a technical career or become discouraged and change majors. Unfortunately, many students fail to apply basic guidelines that could improve their learning efficiency and use of time.

The guidelines that follow should facilitate preclass study, increase student learning in classroom and lab situations, improve performance on exams and reports, and promote student planning and organization. These guidelines are based on the premise that an education should be used by students to learn. Nevertheless, expediency will often dictate that students curtail the depth of their study to meet their obligations to other courses in the form of exams, reports, and homework, all having due dates. Realistically, students should expect to continue learning throughout their technical careers, since they will not be able to learn it all by graduation time. In fact, these students are embarking on a lifetime of learning.

ORGANIZING FOR STUDYING EFFICIENCY Students should find a quiet, comfortable place that can be regularly used for studying. There should be room for textbooks, class notes, calculators, and notebooks, all organized for easy reference. On-campus students may find the library satisfactory, as dormitory rooms may prove distracting.

STUDYING TO LEARN Students should study with a purpose. They should not fall into the trap of reading just to meet assignments due the next day. When reading, they should try not to get lost in the details prevalent in technical material. It is better to skim through the assigned reading first just to see what material the author is developing. It is much better to look at an equation for what it represents than for how it is represented. Students should continue to relate derivations and principles back to the overall intent. A better understanding of technical development will eliminate the confusion that students often feel when they see similar equations and principles applied

The time spent across the desk from a professor can be very valuable to the student. Unfortunately, some students wait until they are in serious trouble in a course before they seek help from their professor. (Courtesy of *Professional Engineer*)

in different courses but expressed in varied notation or applied to unfamiliar problems.

LEARNING FROM OTHER STUDENTS   Students who work together in courses often form lifelong friendships through these associations. Equally important, they assist each other in meeting their daily assignments. It is not unusual for one student to understand a principle or work a problem that another student has struggled with for hours. By explaining problem solutions or basic principles or procedures to one another, students fortify their own understanding of methods and principles. This is important because students are learning to communicate technical information, a quality vital to their professional success. When working with classmates, students sometimes forget that upperclassmen are resources that can be tapped, as are student societies, which often provide tutoring services for their members.

Warning: It is more valuable for students to learn by struggling to understand and then applying course principles to problems than by leaning on others.

TEACHERS ARE RESOURCES FOR LEARNING   Many students shy away from interactions with their teachers other than in classroom or lab situations. This is unfortunate because most teachers are willing to help students, although their busy schedules may not always provide convenient meeting times. Teachers do expect their students to have struggled with their assigned reading, experiment, or homework problem before seeking help. Sometimes they will not give answers, but only advise which step to take next.

LEARNING FROM LECTURES AND LABORATORIES   Students may fail to benefit from classroom instruction and laboratory practice. This is evident when they come to class or lab without having read their assignments or lab instructions; repeat questions that have already been answered or ask questions that show lack of class or lab preparation; sit in the back of a sparsely filled classroom or look out windows, making communication difficult for the instructor; study material from other classes during the lecture; come into class or lab late, or rustle their books five minutes before the end of the class period; or want to turn in homework assignments or reports late or change the dates of scheduled exams.

Students exiting a lecture commonly express their confusion with the lecture content by stating, "Boy, was it snowing in class today." Sometimes this may be caused by an instructor's lack of preparation, but too often it is because students do not prepare in advance. The following suggestions may help; they apply equally well to labs and nontechnical courses.

- First, students should read the assignment or review the laboratory procedures before the class or lab session. When this is done, class lectures or lab sessions have a better chance of offering learning reinforcement rather than confusion. What is not comprehended in preclass study may be understood when presented by the instructor in a different way.
- Note-taking and lecture comprehension may not be compatible. Students taking notes divert their attention from the class or lab presentation to the notes on the blackboard or screen or in their own notebooks. Furious note-taking is a sign of unpreparedness and usually results when students are unsure of the relative merits of the lecture content. A better approach is to copy only that material not in the text or material that reflects the instructor's emphasis. Some students will work in groups, one taking notes while the others listen to the lecture. Then they will gather afterward to review and discuss the notes, making corrections or additions.
- If the classroom or lab situation permits instructor-student interaction, students should take advantage of it. Most instructors will welcome questions when they are thoughtfully posed. Their answers can then give students a glimpse of the methods that capable engineers use to unfold problem solutions or to develop engineering principles. Students may discover that some instructors develop principles, solve problems, or explain laboratory procedures, whereas others expedite the process by skipping intermediate steps of reasoning or equation manipulation that they consider "intuitively obvious." Students who prepare according to the previous recommendations should be able to follow most of these

reasoning or logic jumps; but when they are unable to do so, they should ask questions.

- It is as important to learn why something works as to learn how it is accomplished. Students make a mistake when they search for procedures to solve problems without understanding the basic principles behind them. True, engineers and technologists differ from scientists and mathematicians in that they actually apply fundamental laws, but this is done only after understanding them. Students who engage in "shallow" learning by leaning on procedures find great difficulty in applying the same fundamental principles to problems that are different from those previously encountered.

Homework, Reports, Examinations

One of the most difficult tasks confronting instructors is the assignment of grades. Through performance on reports, homework, and examinations, instructors must assess how well students have satisfied the course requirements. Most instructors will discharge this responsibility in an objective fashion by structuring their grading policy, assigning relative worth percentages to the homework, reports, and examinations. For example, an instructor teaching plant design may feel that students best demonstrate their understanding of course design principles when they apply them to design projects. This same instructor may periodically give exams or homework assignments to stimulate timely learning. It would not be unusual to find the course grading percentages assigned as follows:

Homework	20%
2 exams (20% each)	40%
Design project	40%
Total	100%

In contrast, statics instructors may evaluate their students' knowledge of forces, moments, reactions, and equilibrium by their performance on exams and by having them work numerous homework problems. In addition, they may emphasize a final exam and restructure the grading accordingly, perhaps something like this:

Homework	20%
5 exams (10% each)	50%
Final exam	30%
Total	100%

In other classes, the grading policies may appear more subjective as instructors evaluate their students in class through panel discussions, oral presentations, or individually through oral examinations. Students must demonstrate that they have attained the course objectives to the satisfaction of the instructor, regardless of the evaluation mechanisms used.

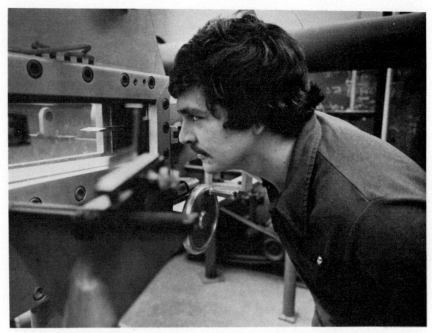

This student is examining the airfoil reaction to the subsonic flow of air past it. (Courtesy of the University of Massachusetts and ASEE)

It is an inescapable fact that students are evaluated best when they "speak the language" of the instructor. And students hear many different languages as they pursue a technical curriculum. In technical courses the symbolic notation may be different from class to class, or one laboratory instructor may demand a different report format than another. One instructor may require all homework to be done in pencil and only on one side of the paper; another may permit homework on both sides. Some instructors will be quite strict in their enforcement of course policies and procedures; others may be more lenient. These differences should not be considered barriers to student achievement but, rather, as a variety of experiences in which students can demonstrate their achievement versatility. Engineering is not standardized throughout the world. To standardize all notation, teaching and evaluation methods, and solution methods would only stifle the creative base for technological development.

Students may not find comfort in the revelation that different learning and evaluation situations are "good for them," for they are struggling to reach their objectives to learn, to make good grades, to demonstrate course accountability, and, ultimately, to graduate. To them, these "experiences" may really appear as "barriers," ones they would just as soon not have to hurdle. Most upper-level students have adjusted to this environment, whereas entering students will probably be frustrated by it. Students may find that by applying the following guidelines, they can better demonstrate their attainment of course objectives. None of these, however, are substitutes for knowing what an instructor expects and responding accordingly.

HOMEWORK GUIDELINES FOR TECHNICAL COURSES When in doubt about what the instructor expects, do the following:

- Work on only one side of the paper, usually $8\frac{1}{2} \times 11$ inches.
- Use "engineering" paper and avoid using paper torn from spiral-bound notebooks.
- Work problems using a no. 2 pencil or a mechanical pencil with F or H lead. This allows erasures to be made.
- Provide a heading for each homework page, including the course number, problem number, student name, and page number.
- Begin each problem with a problem statement. Work the problem in a neat, logical sequence, identifying the information given, the information to be found, the assumptions made, and the principles applied. Printing is preferable, and all intermediate and final answers should be underlined or blocked.
- Use a straightedge for the straight lines in figures, diagrams, and graphs, since it takes about the same time as freehand sketching and is more pleasing to the eye.
- Turn homework in on time, in proper sequence, folded down the middle with identification information—student name, date, course and section number, and problem numbers—listed on the back of the last sheet.

EXAMINATIONS IN TECHNICAL COURSES Quizzes and examinations remain the primary means by which instructors determine course grades. And final course grades will ultimately determine a student's grade-point average (GPA), still the measure of a student's technical ability on graduation, although engineering success cannot always be correlated with the grade-point average.*

On a 4.0 scale, students usually cannot graduate unless they have compiled at least a 2.0 average, the minimum possible C average. And graduate schools may require their prospective students to have a 3.0 GPA (the minimum B average) or better. Consequently, engineering and technology students should try to excel on their course quizzes and exams. The guidelines that follow should enable most students to obtain better grades. These guidelines are concerned with two basic areas: preparing for the exam and taking the exam (see Box 3.3).

---

*J. E. Stice, "Grades and Test Scores: Do They Predict Adult Achievement?" *Journal of Engineering Education*, February 1979, pp. 390–393.

Box 3.3
Do's and Don'ts for Exam Preparation

In preparing for examinations, remember these do's:

- Do spend more time studying material that the instructor emphasizes.
- Do review reading assignments, lecture notes, homework solutions, and quizzes and exams from the same course given in previous years (engineering societies and social fraternities often compile them).
- Do study several days in advance.
- Do study course principles and practice by applying and deriving them.
- Do work problems similar to those expected on the exam.
- Do get a good night's rest before the exam.

Remember these don'ts:

- Don't study "only" that material which the instructor has emphasized. The instructor may check to see if you have completed "all" the assigned work.
- Don't stay up all night preparing for an exam. Technical exams require alert minds.
- Don't forget to bring a charged calculator to the exam, and possibly extra batteries or a charger.
- In open-book exams, don't forget to organize class notes, homework solutions, and other class material for easy reference.
- Don't forget to bring scratch paper and extra pencils.

During the examinations, keep in mind these do's:

- Do sit close to where the instructor usually hands out the exams, usually in the front.
- Do try to relax.
- Do take the time, even though the exam period is limited, to review the exam, noting the relative worth of each problem and any instructions given.
- Do answer the easy questions and solve the easy problems first.
- Do divide your time according to the relative worth of each problem or question.
- Do read each question carefully, making sure that you understand what is to be answered or found. If you do not understand a question, ask the instructor for clarification.
- Do work the exam in pencil and try to be as neat as possible.

- Do organize your problem solutions so that the instructor can follow them.
- Do underline or block in your answers so that the instructor can find them.
- Do keep account of the time left.

Remember these don'ts:

- Don't forget to write your name on the exam.
- Don't panic after reviewing the exam. You may not see all the solutions right away, but some will become evident.
- Don't get lost in the solution of any one problem at the expense of others unless its relative worth merits the extra time; a student's pride can sometimes prove detrimental here.
- Don't forget to specify the units (ft, cm/s) for each answer as you underline or block them in.
- Don't turn in an exam without checking all your answers first, if time is available. Make sure each answer is logical, and also correct spelling errors.
- Don't leave any questions blank, particularly true/false or multiple choice, unless you are penalized for wrong answers. Even if you do not have time to perform the algebraic manipulations involved in a problem, write down how you would solve it if you had the time, assuming that you understand how to do it. Partial credit is often given.
- Don't turn in your exam without checking to see if it is in proper order and that all worksheets have been included.

Those Other Courses

People are stimulated more by activities they either enjoy or consider important. This holds true for students faced with "those other courses"—English, literature, history, political science, economics, sociology, foreign languages. Engineering and technology students can easily correlate technical courses with the prerequisites demanded by their profession; but humanities and social science courses do not always seem to fit into the "master plan." Consequently, students sometimes take a negative attitude toward them.

Humanities and social science courses play an important role in the development of modern engineers, particularly since modern engineering is involved in a technological revolution—one that is actually affecting the rate of social evolution. Social evolution has, in turn, stimulated new relationships between technology and society, placing new constraints on the activities of engineers. These new constraints are causing engineers to act as representatives or intermediaries between industries and governments, between new technologies and communities (where these technologies will be located), and

These young engineers are engaged in planning and budgeting activities, which are essential to most companies. Yet, knowledge of planning and budgeting, like the knowledge gained from courses in humanities and social sciences, is not always perceived as being necessary to the career development of engineering students. (Courtesy of Phillips Petroleum Company)

between management and labor. How, then, can engineers act responsibly in this environment unless they have developed an awareness of societies, including their languages, customs, governments, and basic structures?

Since humanities and social science courses present a slightly different learning environment from technical courses, students may be able to profit from them by heeding the following guidelines.

IDEAS ARE MORE IMPORTANT THAN DETAILS   In courses such as literature, psychology, philosophy, political science, and sociology, ideas are the center of attention. This contrasts with the study of technical details, which absorbs much of a student's study time. Ideas will usually revolve around the way people interact (social customs and government), the way they think (philosophies), and the way they communicate (language). Thus, students will spend considerable time interpreting and communicating the ideas of others, including their instructors. They should take an objective stance as they read their assignments, looking for central themes or concepts. Incidents in the reading material become important when they reinforce these themes. When the central themes are difficult to interpret, students may find that by adopting or studying the point of view of the instructor, they can come to some understanding of the material. Sometimes this understanding will stimulate insights of their own.

IDEAS MUST BE COMMUNICATED   Once ideas are identified, students will be judged by how well they communicate them, whether in examinations, essays, or class discussions. For example, students may be asked to discuss in class the central theme of a poem and describe how the allusions presented in the poem develop the theme. Or they may be asked in an exam to discuss how the religious philosophies of an important historical era influenced the cultural life of that period. Student preparation is crucial if he or she is to impress the instructor. And student preparation may vary with the situation. Take-home essays should be written only after an outline has been prepared and the material to develop the outline has been read, analyzed, collected, and ordered. When faced with class examinations, students may prepare by identifying the key ideas (philosophies, religions, governments, cultures) brought out in the class discussions or reading, along with the supporting details, noting particularly those emphasized by the instructor. If past exams are obtainable, students should review them. Students sometimes find the approach of answering questions that the instructor "may ask" useful. On the examination, students should answer each discussion or essay question only after preparing a brief outline. It is crucial that students make every effort to eliminate misspelled words from their writing, to use correct grammar, and to present neat work to their instructors (typing is preferable).

RELATING IDEAS TO TECHNOLOGY   Students may not improve their grades by relating ideas to technology, but the courses may prove more enjoyable if they do. Consider such questions as "How would cultural philosophies relate to new technological development?" and "How can new technical principles and devices be introduced to various peoples without threatening them?"

PREPARING FOR A PROFESSION

As the engineering profession has matured, a new responsibility has been placed on engineering and technology education: to instill not only a body of knowledge and a set of skills but also a way of thinking, that intangible characteristic called *professionalism*. It is unrealistic to think that a college education can make engineering and technology students true professionals by graduation time. Professionalism comes with experience and time, as young engineers feel their way through their careers, as they make moral and ethical decisions, and as they develop pride in their abilities and continue to improve them as they serve others. Nevertheless, the nurturing of professionalism will begin within the framework of a college education. Educators will play an important role in this growth process, particularly since they provide the first role models for many engineering students.

Educators plant the important seeds of professionalism by the way they lead discussions on controversial technologies and by their promotion of and involvement in technical seminars and conferences. They will promote and support student society membership, sometimes acting as society advisers, and encourage student involvement in extracurricular engineering activities such as plant tours, engineering newsletters, and student service projects. Still, the efforts of educators to instill professionalism in their students will only be successful if the students are willing to nurture the seeds that have been planted.

Engineering students can promote their professional growth by taking advantage of the courses, programs, and organizations afforded them during their college years. Colleges of engineering and technology often sponsor seminars, conferences, or short courses not required for students but which cover current technical topics of interest or display new technologies. These kinds of activities often involve visiting engineers and practical engineering subjects and lead to technical discussions from which students could benefit.

Cooperative Education

The co-op (cooperative) education program is rapidly gaining popularity among students. After their freshman year, students with average grades or better may apply for a work phase in industry as an engineering or technical assistant. This allows them to work with engineers in industry while still in college.

For some the integration of co-op work phases within the normal education framework offers distinct advantages. To begin with, co-op students get a preview of the profession. They are exposed to the industrial environment and how engineers and technologists integrate into this environment. Moreover, co-op students are exposed to companies that they may be interested in working for (and companies have the opportunity to judge prospective employees), and they also have the opportunity to travel and live with co-op students from other colleges. Another advantage is that co-op students earn a salary (more than half what they will be offered on graduation) that can be used to complete their education. And in fact, co-op students receive, on the average, higher salary offers than students graduating without co-op experience. Finally, co-op experiences may motivate students to identify needed educational goals and often provide career encouragement.

Students contemplating a technical career may find the co-op experience just the thing they need to arrive at a career decision. On the other hand, there are some disadvantages to the co-op experience. Co-op work phases last three to six months and may not cover the summer period. Moreover, most co-op students will lengthen their graduation date by a year or more, depending on the number of work phases. Then, too, co-op students may have to work away from their colleges. (This is particularly true for colleges not located in industrial centers.) And co-op students may get out of phase with their curriculum, a problem that students can control by carefully planning their curricula around the work phases and by taking night courses at other colleges near their work, if available. Finally, many co-op students have trouble adjusting from an 8–5 co-op workday to the schedule required in college.

Young co-op engineers sometimes have opportunities to learn engineering operations through "hands-on" experience. (Courtesy of Phillips Petroleum Company)

Society Membership

Another good way students can grow professionally is through active membership in technical societies. Almost every technical society sponsors student chapters if there is a demand for them. Student chapters are often linked to local chapters composed of practicing engineers or technologists, thereby offering students additional opportunities to interact with practicing professionals at the society level.

Student membership is inexpensive, only a fraction of what it will cost when students graduate and enter the profession. Many student societies have membership standards that are not too strict, requiring only that their members be successfully completing the curriculum. Some of the honor societies will impose stricter requirements, such as a minimum grade-point average and recommendations from current members and professors.

In addition to sponsoring field trips, lectures, and local meetings, many societies also sponsor conferences, conventions, and design competitions.

And some student societies actually provide tutoring for their members. Still another advantage of many student societies is that members receive monthly periodicals reviewing their profession. Some societies even publish student journals in which students can publish their own papers following peer review and acceptance. Finally, societies offer students the opportunity to run for student office, direct committees, engage in service projects, and manage budgets.

ENTERING THE PROFESSION

Students approaching their last semester of undergraduate school are faced with two options: either to continue their education via graduate school or to enter the profession. (Actually, there are other options, such as M.B.A. programs, Peace Corps service, and medical school, but engineering students who take these routes are in the minority.) The figures show that most engineering and technology students enter the profession with bachelor's degrees, although more students today are obtaining advanced degrees. This is apparent when one notes that in the late 1970s approximately 70 percent of the engineering graduates were at the bachelor's level, 25 percent at the master's level, and 5 percent at the doctoral level. These figures contrast significantly with those in 1950 when 91 percent were graduating with bachelor's degrees, 8 percent with master's degrees, and only 1 percent with doctorates. Part of this change can be attributed to current industrial attitudes that encourage engineers to obtain advanced degrees. Many companies either sponsor in-house continuing-education programs in cooperation with local institutions or allow their employees time to attend graduate classes or short courses.

There are other pressures for higher education, such as the increase in engineering technology associated with the increase in technological knowledge. Many engineering students find that four or five years of undergraduate study are insufficient and that they need one or more years of graduate study to assimilate the many engineering analytical and design principles to which they were exposed during their undergraduate education.

Then there is the pressure of numbers, and engineers who have advanced degrees may be more visible than those who do not. On the average, engineers with advanced degrees have higher salaries than their counterparts with only bachelor's degrees, and they have greater opportunities for promotion. In addition, more engineers with master's and Ph.D. degrees are moving through the scale of work functions. Whereas Ph.D.'s were once concentrated in either teaching or industrial research, larger numbers now find satisfaction in management, design, development, and manufacturing.

Students who decide to enter the profession immediately upon graduation probably approach the job-hunting period with some trepidation. They may be uneasy about the unknown qualities of the new environment they are about to enter or they may lack confidence in their abilities to "sell themselves" to

The world of engineering has changed considerably in the last few years. Young women have discovered that engineering offers a professional career that is personally rewarding and challenging. (Courtesy of DOW Chemical Company)

prospective employers. These feelings may be further intensified by a poor market for engineers or technologists at graduation time. On a more positive note, students having technical backgrounds have been among the most employable of all college graduates.

There are four major steps to entering the profession: preparing a résumé, obtaining an interview, making a trip to the plant, and making the job decision. Both the student and the employer play a major role in each of these steps. The purpose of the following sections is to review important elements in these steps, elements that will allow students to make the transition from student to practicing engineer or technologist.

The Résumé

A résumé is a short outline of a person's qualifications and experiences that can be used by an employer to assess the "potential" of a prospective employee. It rarely should be more than two pages, and in many cases should be only one page long. Each student who is job hunting should prepare one. It is often expected, even when job application forms must also be filled out.

Students must not underestimate the value of a well-prepared résumé (and well-prepared job application form) in advertising their skills. Because it may provide an individual's first introduction to an industrial representative, the résumé plays an important role in the impressions formed by the interviewer. For these reasons, students may wish to seek professional help when organizing and typing their résumé. Although there are many different résumé formats, most will contain the following elements, although not necessarily in the order presented here.

PERSONAL INFORMATION Students should provide full name, address, and telephone number (both work and home number). Students have the option of providing age, marital status, and number of children.

EDUCATIONAL INFORMATION Students should include major field(s) of engineering and degree(s), minors (if any), specialty areas (if any), grade-point average, and honors (honor societies, scholarships, awards).

JOB INTERESTS This section should contain a brief statement of job interests, but students should be flexible here. Interests that are too specific may limit job opportunities.

EXPERIENCE Students should chronologically outline, from the present to the past, all jobs that demonstrate any engineering experience. Students who have worked summers or in co-op programs can, to their advantage, display real experience. Students who have worked as lab assistants or graders for engineering instructors should also list these experiences. A brief explanation of the responsibilities associated with these jobs might prove fruitful. Students without any engineering experience may wish to describe briefly any jobs that they have had, even though unrelated to engineering. Interviewers are always impressed by students who have worked previously. If students have had no job experience, then they should replace this section with one listing personal qualifications such as motivation and aptitude.

ACTIVITIES Students should list past and present activities that demonstrate a well-rounded background. Involvement in technical societies, any offices held, other student organizations, athletic endeavors, church involvement, and other service activities demonstrate an individual's versatility.

PERSONAL QUALIFICATIONS Students, at their discretion, may list other achievements and honors, plus any hobbies, talents, and personal interests in this section.

REFERENCES Students should provide the names, addresses, and telephone numbers of several references who can testify to their qualifications and character. However, students must always remember to ask permission of anyone whose name is to be listed.

A résumé's worth in securing a job should not be overestimated, since the most important measuring tool is probably the interview. If the résumé helps students secure more interviews, then it is fulfilling its purpose.

The Interview

Many students will interview with industrial representatives in the job placement office on campus. Others may secure interviews at off-campus industrial sites, often in other towns or states. Résumés, letters of application, and personal contacts may open the door to interviews, but it is absolutely crucial that they be done with great care. If not, off-campus interviews will be difficult to obtain and on-campus interviews, more freely obtained, may be seriously hindered, even before the student walks into the interviewer's office. For example, forms filled out sloppily may lead the interviewer to make judgments about the quality of the applicant's engineering work.

Once a student enters the interviewer's office, he or she will be scrutinized for about ten to twenty minutes, on the average. This means that only ten to twenty minutes are available to impress the interviewer. Thus, the first step is to make a good "second" impression by presenting the professional look often conveyed by dress, posture, a firm handshake, and a steady eye. (The first impression has already been left by the résumés, application forms, or personal contacts.) Students should be well dressed and well groomed; this conveys to the interviewer that the student is serious about the interview.

Next the student should assume a relaxed posture and look directly at the interviewer. Nervous habits such as smoking, cracking knuckles, or gum chewing should be avoided as the student waits for the interviewer to ask questions. It is much better for students to thoughtfully respond to an interviewer's question than to ask unsolicited ones. Let the interviewer set the tone of the discussion first; there will be opportunities later for a student to ask questions. Remember, the interviewer is interested in a student's judgment, poise, and self-control.

There is nothing wrong with trying to prepare for the interview. Learn something about the company and the company representative, if possible. There will usually be some company literature to review in the placement office. Practice by responding to questions that you think might be asked. Although it is difficult to predict the kinds of questions that may be asked engineering applicants, those that follow are representative.

- Would you work nights and weekends, if necessary, to complete an assignment?
- Where do you picture yourself five years from now in terms of position and income?
- Since you are a chemical (civil, electrical, . . .) engineer, how would you apply this chemical (civil, electrical, . . .) principle to this problem?
- Do you prefer working on individual engineering assignments or with other engineers and technologists on cooperative projects?
- Would you rather delegate a responsibility or do it yourself?
- How do you feel about company transfers?
- How would you describe your college education and your engineering achievements?

- What do you consider your best characteristic?
- What personality trait do you need to improve?

After the interviewer has asked a number of questions, time may be left for the applicant to ask questions. Often this depends on the tone of the interview and the encouragement of the interviewer. Questions that show sincere interest in the company and a knowledge of its products may impress the interviewer.

Finally, the last favorable impression that an applicant can leave is a firm handshake and words that demonstrate real interest in the job. The interview, it is hoped, will be successful enough to get the student applicant invited to the company for a visit, sometimes called the plant trip.

The Plant Trip

Several weeks after the interview, student applicants will either receive a letter of rejection, be offered a job with or without an accompanying plant trip, or be offered a plant trip without a firm job commitment. But whenever plant trips are offered, students can assume that their chances of being offered a job have increased. Plant trips involve considerable expense to the company plus a commitment of personnel to conduct secondary interviews and guided tours; they are not offered unless the company is definitely interested in the applicant. The applicant should feel good about this opportunity to visit the company, for he or she has passed the first test. Now the applicant will be scrutinized even closer and over a longer period of time, but perhaps in a less formal way.

After reporting to the company at the appointed time, dressed in a professional manner, the applicant will be met by a company representative, possibly a staff engineer from one of the areas to which the applicant seems best suited. (If the applicant has had to travel a significant distance, the applicant will probably travel by air at the company's expense, stay in a motel or hotel, and be escorted around by company representatives.) Then the applicant will be led through a series of interviews with various personnel, some formal and some informal. From these interviews and informal conversations, the company will receive several opinions about the applicant's technical ability, compatibility with others, and level of personal responsibility.

At the same time, the applicant has the opportunity to examine the company. There are many things the applicant may wish to assess.

COMPANY WORKING CONDITIONS Are the facilities comfortable? Do the engineers work in large bays, or do they have individual or shared offices? Has the company sufficient numbers of technical aides, computers, test equipment, and so forth?

COMPANY MORALE Do the engineers, technologists, and technicians seem happy and enthusiastic about what they are doing? Is the environment

This is the office of a technical professional. It is not unusual for engineers to divide their time between comfortable offices like this one and rough jobs in the field. (Courtesy of Phillips Petroleum Company)

stale or dynamic? Is there some flexibility in the working hours? What kind of relationships seem to exist between the supervisors and those they supervise?

COMPANY BENEFITS   What kinds of retirement and insurance plans are provided? How many paid days of vacation? Can the employees take their vacation at their discretion? Is there a company stock plan?

OPPORTUNITIES FOR GROWTH   Does the company encourage and promote continuing education? Are there opportunities to move within the company to better and more challenging positions? Does the company encourage its employees to attend seminars, conferences, and short courses? Does the company provide in-house training? Are the younger employees given responsible assignments?

In most cases, students won't be offered a job at the conclusion of the plant trip, but they won't receive a job rejection either. It usually takes several weeks for a company to assess an applicant's potential in comparison with others who have interviewed for the same or similar jobs. Later, if the applicant receives a personal phone call or telegram from the company representative, chances are good that a job is being offered. If the applicant receives a letter, it may be an offer or a rejection. Once an applicant receives one or more job offers, he or she is faced with making a job decision—no easy task.

Most major companies invite guest experts to present lectures on technical topics. Here a visiting professor discusses advances in polymerization processes with engineers and scientists. (Courtesy of Phillips Petroleum Company)

The Job Decision

If only one job offer is received, then the decision is easy to make, unless the offer is totally unacceptable. It is hoped, however, that students will have the opportunity to choose between several job offers. In this situation, there are several considerations that can affect the decision finally made.

ECONOMICS   The job with the highest salary may not always provide the best economic picture. The cost of living varies significantly from region to region, depending, among other things, on local and state income, sales, and property taxes. Insurance rates may be higher in some areas than in others, and energy and utility costs may vary considerably. Housing in larger cities tends to be more expensive than in rural areas. Variations in the cost of these items can affect the net worth of a salary by thousands of dollars. The plant trip, described earlier, provides an opportunity for applicants to investigate the economic circumstances of the community, area, and state in which the company is located.

Young engineers can benefit from the experience of more mature engineers. Much of engineering relies on intuition gained from years of on-the-job experience—something that only time can teach. (Courtesy of DOW Chemical Company)

QUALITY OF LIFE   What constitutes the "good life" will vary from applicant to applicant. For applicants with families, decisions may be influenced by the quality of schools, neighborhoods, churches, and recreational facilities. Other applicants may be interested in the availability of outdoor activities; still others may desire areas in which cultural, educational, and athletic activities thrive.

THE WORK ENVIRONMENT   The job environment is one of the key considerations for any prospective employee. The plant trip should have answered most of the questions that applicants have about the environment in which they will work and about the potential for job advancement, self-improvement, and overall job satisfaction.

Once students make their job decision, they should get in touch with the representatives of those companies that made offers, either by mail or telephone, and inform them of their decision. Students should be concerned but not overly anxious about the wisdom of their decision. If students discover later that they have made a mistake, they should not look upon it as a disaster. Instead, they should learn from the job what they can by contributing as much as possible while looking for a more suitable job. Most engineers and technologists have several jobs throughout their careers.

THE PROFESSIONAL YEARS

The career of an engineer or technologist is challenging and sometimes unpredictable. Students entering the profession will soon find this to be true. In a very short time, most of their knowledge and skills will be tested in an environment different from college in many ways.

Some young engineers and technologists will be thrust into programs or projects demanding their immediate attention. They will be expected to perform technical work consistent with the skills and abilities they are assessed as having.

Other organizations may introduce their new employees to the system more slowly by providing a significant orientation and training period; some companies invest up to a full year in the initial training of their new employees. Because they are eager to engage in "real" engineering, new engineers may find this period quite frustrating; yet it is an important period of integration for the new engineer or technologist.

Most engineers and technologists find that they change employers several times; thus, students should look on their first job after graduation as an opportunity to learn as much as possible. If they like the company and the company likes them, so much the better. But if not, it is not a disaster; other jobs will become available.

A technical education does not stop after graduation, although its emphasis may change. Engineers and technologists must keep up with advances in their technical fields. They should follow the developments in the technical journals of their own field as well as the overall technological developments in other fields through magazines like *Scientific American* or *Science News*.

Employer Expectations

Employers usually expect their employees to exhibit good work ethics, to be technically competent, to work cooperatively with their co-workers, to take an interest in their jobs, to be loyal to the company, and to be active in community affairs and organizations.

New employees are expected to have fundamental technical knowledge, but not necessarily to be abreast of the latest company innovations, particularly in the specialty areas to which they are now assigned. But their employers do expect them to learn as rapidly as possible. In fact, how rapidly they learn about these new specialty areas is an important factor to employers when the time comes to judge abilities and potentials.

All employers are flattered and pleased by employees who work hard and take a real interest in their jobs. This may require concentrated effort at times when the engineering tasks are not exciting or challenging. The excitement of problem solving and design innovation will not surround every engineering task. There is a certain amount of dull repetition in any engineering job.

Employers are also impressed by engineers with good ideas, even though resistance to their ideas may exist; this resistance is called organizational inertia. The ideas of new employees, who are often in the best position to see

Young engineers participate in activities to advance the communities in which they live. One community organization that opens avenues for service is the Jaycees. (Courtesy of Phillips Petroleum Company)

Here a young engineer acts as leader of a scouting group, helping youngsters learn about canoeing and the outdoors. (Courtesy of Phillips Petroleum Company)

areas for improvement, are frequently viewed with skepticism. Those with good ideas will have to discover for themselves ways to introduce their ideas into the organization. This may require creative steering around the inertial roadblocks that always seem to exist.

Employers have every right to expect loyalty from their employees. And engineering professionals have the responsibility to provide loyal service. New ideas and designs conceived within or stimulated by the company work environment are the proprietary property of the company and should be protected from infringement by others, in much the same way as medical doctors honor the confidences of their patients. Of course, all employees should recognize that even loyalty has its limitations and that the goal of engineering is to provide products, devices, and methods for the "good" of society.

In addition, engineers and technologists should avoid getting into dead-end jobs. It is not disloyal to avoid a job that becomes too routine. Those in routine jobs are often overlooked by the company and thus are likely to become obsolete. Clearly, this is not good for the company or the employee. Engineers or technologists who find themselves gravitating toward this situation should take steps to transfer within the company, or, if necessary, start looking outside the company.

A company is proud of its employees' participation in community affairs and organizations. These activities can only help an employee's chances for advancement. In addition, it will expose engineers to society, broaden their perspective, and enable them to more comfortably interact with others.

Succeeding with People

Engineering is basically a group effort. As a result, an organization's success often depends on how well the individual members of a group work together. Groups can increase productivity and effectiveness if organized correctly and if the individual members understand how to contribute in a group situation. In contrast, a group of "individuals" may end up producing a product less desirable than each individual might have produced alone. This phenomenon is embodied in the saying, "A hippopotamus is a horse that was produced by a committee."

A technical education is oriented toward teaching prospective technologists or engineers how to function as individuals. Little time is spent developing a student's interpersonal skills; yet practicing professionals know that their effectiveness is measured, in part, by how well they function with other individuals, whether in groups or on a one-to-one basis. The following suggestions should help one's interpersonal skills.

PEOPLE ARE INDIVIDUALS, TOO   One of the biggest mistakes we make is in assuming that everyone thinks as we do. Fortunately, they do not. People think differently, act differently, and perceive things differently. We must expect these differences to exist in people and continue to see their view of the world as we work with them. It is these differences that often provide the environment we need for engineering innovation.

PLACING PEOPLE ON THE OFFENSIVE   The simple posing of a sentence or question can set the tone of a one-to-one conversation. Sentences and questions that are dictatory or challenging often force others to assume a defensive posture. In contrast, sentences and questions that solicit opinions and advice place people in a constructive or offensive posture. In most cases people want to contribute and be respected for their ideas. It is therefore helpful to ask questions such as, "What do you think?" and "How can we . . . ?"

PATIENCE REALLY IS A VIRTUE   When working with others, we must be prepared to listen, sometimes at the expense of expediency. This requires patience, emotional control, and sometimes an effort of concentration.

PREPARING FIRST   Engineers who are well prepared are less likely to interact poorly with people and can maintain a sincere, objective viewpoint rather than a subjective one, which may lead to an emotional situation.

PROBLEMS

3.1   Tabulate your activities for a week, listing the number of hours spent in recreation, studying, eating, sleeping, traveling, attending class, and working. Use this list to evaluate how well you are using your time. If necessary, develop a new schedule that will allow you to allocate your time more efficiently. How many hours are you spending in study for each class?

3.2   List all your characteristics that you believe are consistent with those required by engineers and technologists. Compare these with the list in Box 3.1.

3.3   Review the educational goals in Box 3.2. Write down any other goals that you believe should be added to this list.

3.4   Review entirely the required courses in your technical curriculum. Which two prerequisites influence the greatest number of courses?

3.5   Write down evaluation procedures by which you will be judged in each of your classes. Include the percentages assigned to each exam and homework. Are you not performing in some of the grading areas? Do you understand how you are being graded?

3.6   Determine which humanities and social science courses are acceptable toward meeting your curriculum requirements.

3.7   Determine the minimum requirements to participate in co-op programs at your institution. What are the average salaries being paid to co-op students?

3.8   Interview a graduate student in your branch of engineering (or possible technology discipline). Determine why he or she is attending graduate school.

3.9   Prepare a résumé based on your education, talents, and job experiences up to the present. Identify those areas that are conspicuous by their absence. Lay out a plan to erase these deficiencies.

3.10   Visit the placement office during the interviewing season. Observe the actions of senior engineering students and the industrial representatives. List those things that you consider noteworthy.

3.11   Interview a practicing engineer or technologist who has recently graduated and has been working for less than a year. Write down his or her impressions of the first few months in industry.

The Profession

CONTENTS

Professionalism: Characteristics and
    Responsibilities
Cases in Ethical Studies

Engineering became a profession only when it was recognized as such by society. In turn, engineering professionals have inherited a responsibility to society that can never be treated lightly: to utilize materials and technology in the service of humankind. One of the major functions of the profession is to ensure that this occurs.

PROFESSIONALISM: CHARACTERISTICS AND RESPONSIBILITIES

Herbert Hoover, the thirty-first president of the United States and a self-acknowledged engineer, made the following statement about engineering in his memoirs:

It is a great profession. There is the fascination of watching a figment of the imagination emerge through the aid of science to a plan on paper. Then it moves to realization in stone or metal or energy. Then it brings jobs and homes to men. Then it elevates the standards of living and adds to the comforts of life. That is the engineer's high privilege.

The great liability of the engineer compared to men of other professions is that his works are out in the open where all can see them. His acts, step by step, are in hard substance. He cannot bury his mistakes in the grave like the doctors. He cannot argue them into thin air or blame the judge like the lawyers. He cannot, like the architect, cover his failures with trees and vines. He cannot, like the politicians, screen his shortcomings by blaming his opponents and hope that the people will forget. The engineer simply cannot deny that he did it. If his works do not work, he is damned. That is the phantasmagoria that haunts his nights and dogs his days. He comes from the job at the end of the day resolved to calculate it again. He wakes in the morning. All day he shivers at the thought of the bugs which will inevitably appear to jolt its smooth consummation.

On the other hand, unlike the doctor, his is not a life among the weak. Unlike the soldier, destruction is not his purpose. Unlike the lawyer, quarrels are not his daily bread. To the engineer falls the job of clothing the bare bones of science with life, comfort, and hope. . . .

The engineer performs many public functions from which he gets only philosophical satisfaction. Most people do not know it, but he is an economic and social force. Every time he discovers a new application of science, thereby creating a new industry, providing new jobs, adding to the standards of living, he also disturbs every thing that is. New laws and regulations have to be made and new sorts of wickedness curbed. . . . But the engineer himself looks back at the unending stream of goodness which flows from his success with satisfaction that few professions may know.*

When did engineering gain its status as a profession? President Hoover obviously considered it a profession decades ago, yet one cannot point to a specific year when engineering was officially recognized as a profession because there is no formal procedure for bestowing this title. Although we have traced some of the initial professional characteristics of engineering back to antiquity, engineering became a profession only when it was recognized as such by society.

What are the characteristics of learned professions? Reviewing the professions of law and medicine, we can identify certain professional characteristics common to both.

- Recognition as a profession by society and by its practitioners.
- Education, knowledge, and skills in certain areas that exceed those of the general public.
- An attitude of service to the client and loyalty to the employer, respecting and protecting the confidences shared with them.
- Recognition by law.
- Recognition of the need to remain professionally competent by participation in organized societies and continuing education.
- Adherence to specific codes of ethical conduct.

Without a doubt, engineering is recognized, both by its practitioners and by society, as a profession. Although there are some groups and individuals who call themselves engineers while meeting few of the professional characteristics listed previously, their number is small.

Education, Knowledge, and Skills

In the area of education there are distinct differences between engineering, law, and medicine. Almost all practicing engineers and technologists have a

---

*From *Memoirs of Herbert Hoover, Vol. 1, Years of Adventure,* by Herbert Hoover. Copyright 1951 by Macmillan Publishing Co., Inc. Reprinted by permission of the Herbert Hoover Library, West Branch, Iowa.

When does an engineer become a professional? At graduation? Later? Perhaps it is an evolving process, a more mature way of thinking, more pride in one's work, and a concern for others. (Courtesy of The Standard Oil Company)

college education with at least a bachelor's degree, and many have continued their study through the master's and Ph.D. levels. In contrast, lawyers must attend a three-year law school, usually after graduation from college, and medical doctors will pursue even more extensive education beyond college, requiring at least four years of medical training and internship in a recognized medical school. Those specializing in certain areas of medicine, such as surgery and obstetrics, will require several years more of training.

Because engineers can practice their profession with an undergraduate degree, their college course requirements are considerably different from those of the prospective law or medical student. Engineering students will engage in four or five years of intense technical study, in particular engineering areas supported by studies in the maths and sciences. Study in the humanities and social sciences will be important but somewhat limited. On graduation, engineering and technology students will usually receive a bachelor of science (B.S.) degree in their particular field of study, although there are other degrees granted, such as the bachelor of engineering or engineering technology degrees. In contrast, premed and prelaw students can qualify to take entrance exams to law and medical schools after three years of college study in broader areas. Today, most prelaw and premed students are graduating with a bachelor of arts (B.A.) degree before entering their professional school.

Recognition by Law

Recognition by law is usually found in the licensing or registration procedures of most states. There are close to 500 professional and nonprofessional occupations currently licensed by one or more states, and the trend is toward licensing even more occupations. The licensing of professions has become a means of protecting the public and has also proved to be a protection for the professional person by lending status to those who are licensed and by elevating standards of performance and compensation.

The licensing agencies of most states have the following duties:

- To examine the credentials of applicants and determine whether they have sufficient education and experience to meet the licensing requirements.
- To review the current standards of schools in judging their acceptability as training institutions.
- To direct the examinations used to judge the qualifications of prospective licensees.
- To grant licenses.
- To make regulations governing the practice standards, investigate charges of violations, and conduct hearings when necessary. In those cases where violations are proved, licenses may be suspended or revoked.

The licensing boards of professions such as law, medicine, and engineering are usually composed of licensed practitioners from that profession. In many states the authority and responsibility for licensing prospective attorneys at law are delegated to the respective bar associations. Thus, candidates for law practice will have to pass the bar exam to gain membership in the bar association.

Wyoming enacted the first law regulating engineering in 1907; by 1950 all states had such laws. Today we have more than 400 000 registered professional engineers, a large percentage of the estimated 1.1 million engineers now practicing.

Among the registered engineers, we find various levels of education and experience. Many do not have a college engineering degree but have qualified for registration by their years of experience or by their established eminence. In many states, though, this is no longer allowed and an engineering degree from an acceptable institution is a prerequisite for registration. For most new engineers, the steps to becoming a registered professional engineer are as follows:

1. Graduate with an engineering degree from an institution acceptable to the registration board of the state.
2. After graduation and if twenty-one years or older, successfully pass a written examination that certifies the engineer as an Engineer-In-Training (EIT). For most states this fundamental examination is developed by the National Council of Engineering

Examiners (NCEE) and covers basic theory and engineering fundamentals over an eight-hour testing period. It may or may not be an open-book exam. What constitutes passing will then be determined by the state registration board.

3. Spend at least four years practicing engineering and gaining professional experience and be at least twenty-five years of age before submitting the application for registration, including the names of references, some of whom must be licensed engineers.

4. Take an eight-hour open-book exam on the principles and practice of engineering to demonstrate your ability to apply engineering principles and judgment to the solution of problems normally encountered in practice. This examination is also developed and graded by the NCEE, with the registration board again determining what constitutes passing.

On completing these steps and paying the necessary fees, the engineer will be registered in the state as a professional engineer (P.E.) and will now have the authority to practice that branch of engineering that interacts directly with the public—public works.

All engineers are not engaged in public works. Many industries employing large numbers of engineers have products that are not classified as public works, such as aircraft companies, computer companies, and research laboratories. Large numbers of engineers in their employ are not registered, although there seems to be a trend to have more of their key designers registered to afford some protection to the company.

It should be noted that few engineering technologists are registered as licensed engineers. There has been a great reluctance among state boards to recognize technology degrees as sufficient for public engineering practice.

Why License?

The following true story is offered as an example of why licensing procedures for engineers are necessary.* The applicant requested registration based on practical experience, classifying himself as an "amusement engineer." He noted that he had designed the roller coasters in some of our most popular amusement parks. In the oral exam given him, the following questions and answers occurred. Needless to say, he did not receive his license.

QUESTION: How do you calculate stresses in the structure?

ANSWER: I know from experience how large to make all the pieces.

QUESTION: What calculations do you make?

ANSWER: I have a formula in my office for the force of a car going over a curve.

QUESTION: Do you employ any other formulas?

ANSWER: No, sir. That is the only one I need.

*Reprinted by permission from *How to Become a Professional Engineer*, by J. D. Constance. 3rd ed. (New York: McGraw-Hill, 1978).

QUESTION: What factor of safety do you use?

ANSWER: About 90 percent.

QUESTION: Do you mean that your factor of safety is less than unity?

ANSWER: Yes, sir, less than unity.

QUESTION: Why do you do that?

ANSWER: To give the customers a bigger thrill. They have to hold on when they go over a curve.

QUESTION: Isn't that rather dangerous?

ANSWER: No! When a car is going fast, you don't need 100 percent safety. You get across before anything happens.

Service and Loyalty

Engineering is intended to be a profession that serves society by providing technology, products, devices, structures, and procedures that are beneficial. In view of some of our environmental problems, it is obvious that we haven't always been successful. Still, service is a key word, although most engineers will not be able to serve the public on a one-to-one basis as do lawyers and doctors.

The majority of engineers and technologists will work in technical organizations of both small and large companies and interact mostly with other engineers, technicians, technologists, managers, equipment vendors, and other company personnel. Thus, service is rendered by the employee to the company as employer.

Companies that invest a considerable amount of time and money in the training and compensation of their employees expect them to serve to the best of their ability. Loyalty will play an important part in the quality of this service. Since many companies are engaged in competitive practice or in work vital to the security of the nation, engineers and technologists will ultimately develop or come in contact with information or products that must be protected in a proprietary manner.

Professional Competence

One of the greatest challenges that engineers and technologists face is to remain technically competent throughout their career. On graduation from college, the burden of this responsibility will fall on the individual. Ideally, the college education will have implanted the ability to "learn how to learn."

Fortunately, the engineer can look to the engineering profession for help. Like all true learned professions, engineering is composed of societies that promote professional development, including continuing education. These societies will generally be organized at the national, state, and local levels. They will sponsor technical and professional meetings at all levels, publish technical journals, sponsor short courses, and work to improve the standards of engineering education.

There are nearly 200 technical societies or related groups listed in the directory of the Engineers Joint Council (EJC). Among the first major soci-

---

Box 4.1
Founder Societies

ASCE—American Society of Civil Engineers (1852)
AIME—American Institute of Mining, Metallurgical, and Petroleum
  Engineers (1871)
ASME—American Society of Mechanical Engineers (1880)
IEEE—Institute of Electrical and Electronics Engineers (1884)
AIChE—American Institute of Chemical Engineers (1908)

---

eties are the five known as the Founder Societies, which account for a combined membership of approximately 400 000 engineers. The Founder Societies and their dates of organization are listed in Box 4.1.

The Accreditation Board for Engineering and Technology, Inc. (ABET) has been particularly important to students by discharging the responsibility for the accreditation of engineering and technology curricula and for the formulation of the Canons of Ethics. It is composed of participants from fourteen societies (the five Founder Societies and the nine listed in Box 4.2).

Code of Ethics

There is no one code of ethics subscribed to by all engineering societies, many of which have their own. But the code of ethics most widely accepted is probably that of the National Society of Professional Engineers (NSPE). It appears in Box 4.4 at the end of this chapter. By comparing this code with the Fundamental Canons of the Code of Ethics of the Accreditation Board for

---

Box 4.2
Engineering Societies

AIAA—American Institute of Aeronautics and Astronautics (1932)
AIIE—American Institute of Industrial Engineers (1948)
ANS—American Nuclear Society (1954)
ASAE—American Society of Agricultural Engineers (1907)
ASEE—American Society for Engineering Education (1893)
NCEE—National Council of Engineering Examiners (1920)
NICE—National Institute of Ceramic Engineers (1938)
NSPE—National Society of Professional Engineers (1934)
SAE—Society of Automotive Engineers (1905)

Engineering and Technology, Box 4.3, we see similarities that demonstrate the pride engineers have in their profession. It is this kind of pride that makes engineering a profession.

It is difficult to comprehend the professional decisions that can confront practicing engineers and technologists. Ethical decisions are not always black and white, even when codes of ethics are readily available to guide these decisions. To demonstrate this point, two case studies in ethics are presented in the following sections.

CASES IN ETHICAL STUDIES

The following case studies are based on data submitted to the NSPE Board of Ethical Review and do not necessarily represent all the pertinent facts when applied to a specific case. The opinions rendered are for educational purposes only and should not be construed as expressing any opinion on the ethics of specific individuals. Where the word "we" appears, it should be construed as the opinion of the review board.

We will first consider the facts and questions for two cases in ethical studies. We will then present the board's discussion comments and the board's final conclusions. Note that the NSPE Code of Ethics, which is often referenced in these studies, is found in Box 4.4 at the end of this chapter. Also note that these Board of Ethical Review case studies come from the *Professional Engineer,* published periodically by NSPE.

Case 1: Value Engineering—Contingency Fee

FACTS   John Doe, P.E., a principal in a consulting engineering firm, attended a public meeting of a township board of supervisors that had under consideration a water pollution control project with an estimated construction cost of $7 million. Doe presented a so-called cost-saving plan to the supervisors under which his firm would work with the engineering firm retained for the project to find "cost-saving" methods to enable the township to proceed with the project and thereby not lose the federal funding share because of the township's difficulty in financing its share of the project.

Doe further advised the supervisors that his proposal stipulated providing his "cost-saving" services on the basis of being paid 10 percent of the savings, but that his firm would not be paid any amount if it did not achieve a reduction in the construction cost. Doe added that his firm's value engineering approach would be based on an analysis of the plans and specifications prepared by the design firm and that his operation would not require that the design firm be displaced.

QUESTION   Were Doe's presentation and offer consistent with the Code of Ethics?

Box 4.3
Fundamental Canons, ABET Code of Ethics*

1. Engineers shall hold paramount the safety, health and welfare of the public in the performance of their professional duties.
2. Engineers shall perform services only in the areas of their competence.
3. Engineers shall issue public statements only in an objective and truthful manner.
4. Engineers shall act in professional matters for each employer or client as faithful agents or trustees and shall avoid conflicts of interest.
5. Engineers shall build their professional reputation on the merit of their services and shall not compete unfairly with others.
6. Engineers shall act in such a manner as to uphold and enhance the honor, integrity and dignity of the profession.
7. Engineers shall continue their professional development throughout their careers and shall provide opportunities for the professional development of those engineers under their supervision.

*Reprinted courtesy of Accreditation Board for Engineering and Technology, approved by the Board of Directors, October 5, 1977.

REFERENCES *Code of Ethics:* Section III, 7(a)—"Engineers shall not request, propose, or accept a professional commission on a contingent basis under circumstances in which their professional judgment may be compromised."

Case 2: Name of Nonengineers in Professional Corporation

FACTS Two individuals, John Adams and Samuel Boone, formed a consulting engineering firm. In public announcements, Adams and Boone indicated that the firm was to be known as Adams, Boone & Associates. Adams is a registered professional engineer. Boone is not a registered professional engineer but is a certified engineering technician. Because Boone was not a registered professional engineer, the use of Boone's name in the firm was discontinued at the insistence of the state board of engineering registration. A professional corporation was then formed for the firm using the name Adams & Boone, P.C. To meet the requirement of state law that only registered persons be shareholders of a professional corporation, Adams was the sole shareholder. The firm's letterhead indicates that it is "A Consulting Engineering Firm," and Adams is listed on the letterhead as John Adams, P.E., President, and Boone is listed on the letterhead as Samuel Boone, C.E.T., Executive Vice-President.

Both Adams and Boone are full-time employees of the professional corporation and both have contact with the firm's clients in carrying out the firm's consulting engineering business. According to state law, an employee or subordinate of a registered person is exempt from registration, provided his or her duties do not include responsible charge of engineering.

QUESTION May Adams ethically associate with Boone in the manner indicated?

REFERENCES *Code of Ethics:* Section II, 1(d)—"Engineers shall not permit the use of their name or firm name nor associate in business ventures with any person or firm which they have reason to believe is engaging in fraudulent or dishonest business or professional practices."

Section III, 9(a)—"Engineers shall conform with registration laws in the practice of engineering."

Case 1: Discussion and Conclusions

DISCUSSION In an earlier case, call it case A, we decided what, on the face of it, appears to be an almost identical issue, concluding that it was ethical for a firm to offer clients in the industrial field a value engineering service with payment based on a percentage of the savings achieved in the design or production mechanisms of the client. In that case the percentage of savings to be paid the engineering firm would be based on negotiation prior to the rendition of the value engineering service and with the understanding that the *client would have the unilateral right to not utilize the* proposed changes. It is important to note that we based our conclusion, in part, on the understanding that under the prevailing facts the client was informed and competent to judge the proposed changes. Consequently, the engineering firm would have the incentive to avoid compromising its professional judgment.

Subsequently, we dealt with another related case, call it case B, in which a private engineering firm sought a contract to prepare an independent design for a major bridge to compare with an in-house design. The engineering firm offered to provide its services on the basis that it would not be paid if its design did not save the state at least 5 percent of the construction cost of the in-house design.

In that situation we reached a contrary result and held the procedure to be in conflict with the Code of Ethics because the engineering firm "may be tempted to specify an inferior design concept and materials to produce a lower construction cost" (in order to secure its fee). But the discussion in case B recognized that there was an arguable view to reach a contrary result because of the possible benefit to the public if a substantial saving of money could be realized without the sacrifice of safety or quality. In rejecting that arguable premise, however, we said that on its face, Section III, 7(a), of the Code of Ethics bars contingency contracts when professional judgment may be compromised. It appears that contingency contracts, when used as a device for promoting or securing a professional commission, might fit into this category.

If the differing results reached in the two previously decided cases can be reconciled, the rationalization would be that in case A the engineering firm was offering a value engineering service on a contingency basis in general, whereas in case B we were dealing with a very specific project directly related to the public health and safety. Further, as previously observed, in case A, we commented that the client was informed and would have the capability to protect its own interests by careful review, with right of rejection, of any proposed changes to achieve a saving.

The facts before us are more nearly akin to those set forth in case B, pertaining to a specific project clearly related to the public health and safety. Even more urgent, the facts before us indicate a relatively small community as the client; and we may assume, on the basis of practical experience, that such a client would not likely have the expertise of its own to judge and be in a position to recognize if the changes in the design as proposed by Doe's firm would entail dangers or a serious loss of quality for the project. True, the township could call upon the original design firm for comment and recommendation before acting favorably on the changes in the design intended to effect a construction cost reduction. But in that event, the township supervisors would possibly be confronted with opposing opinions of two engineering firms and not be in a position to make a qualified judgment.

CONCLUSIONS   Doe's presentation and offer were not consistent with the Code of Ethics.

Case 2: Discussion and Conclusions

DISCUSSION   We first note in explanation that all states now have "professional association" (P.A.) or "professional corporation" (P.C.) laws that in both instances permit the various professions to operate as corporations for tax purposes. The restrictions and conditions of these laws vary in detail regarding stock ownership, who may be officers or directors, the extent of practice in a particular profession, and restrictions on transfer of stock.

The particular state law in question permits the practice of the various professions under the "professional corporation" form and, presumably, on the basis of the facts submitted, does not require that all officers of the P.C.

be registered in the particular profession. The designation of Boone as executive vice-president clearly implies that he is an officer of the corporation, even though not a shareholder. We further note that the applicable state law provides that a P.C. ". . . may render professional service only through officers, employees, and agents who are themselves duly licensed and otherwise legally authorized to render professional services within this state." But an "employee" not licensed in the particular profession may perform related services under the direct supervision and control of an officer, agent, or other employee who is licensed to render professional service on behalf of the corporation.

Applying the conditions of the law in the state in question to this case, our concern is not with the legality of the form of practice or name of Adams & Boone, P.C. That aspect is a legal issue for determination by the state legal authorities. Rather, our jurisdiction is to determine if the name of a nonengineer on the letterhead in this setting offends Section II, 1(d). There can be no question that Boone is not legally qualified to render the professional services for which the association is intended.

It is not uncommon for engineering firms to list the names of key officers or employees on the letterhead, usually followed by a designation of the individual's professional field, such as P.E. for professional engineer, L.S. for land surveyor, AIA for an architect, CPA for an accountant, and others. We do not perceive of any ethical problem in that type of listing because the public is not then misled into believing that all those listed are professional engineers.

There are, of course, other variations for listing of names, such as showing P.E. after some names with no designation after other names. Again, it would be reasonably clear to the public that those without a designation are not professional engineers. A form of listing that would create an ethical concern would be a listing of names without any designations, in which case there would be an implication that all those listed are engineers if the letterhead indicated the firm to be one offering professional engineering services. The potential problem may be avoided altogether by not listing any individual names on the printed letterhead. In that case the individual signing a letter or other document would show under his or her signature the firm title and whatever professional designation might be appropriate.

Another aspect of this case is the meaning of "associate" in Section II, 1(d), within the context of these facts. We believe that "associate" is usually intended to refer to relations with other firms or clients; that is, that the thrust is external rather than internal.

A related facet of "associate" in this case, however, is the degree to which Boone's contacts with clients or prospective clients may be within the bounds of II, 1(d). We dealt with a related situation in a case in which a nonengineer was employed by an engineering firm to solicit engineering contracts for the firm. We held then, and repeat now, the opinion that it is permissible for a firm to employ a nonengineer as a representative to solicit engineering service contracts, provided, however, that the nonengineer representative may not enter into discussions of engineering aspects of an actual or proposed project with a prospective client. That function may be performed only by registered engineers. This comment would apply in this case to Boone's contacts on behalf of Adams & Boone, P.C.

CONCLUSION   Adams may ethically associate with Boone in the manner indicated.

Box 4.4
NSPE Code of Ethics for Engineers*

NATIONAL SOCIETY OF PROFESSIONAL ENGINEERS FOUNDED 1934

Preamble

Engineering is an important and learned profession. The members of the profession recognize that their work has a direct and vital impact on the quality of life for all people. Accordingly, the services provided by engineers require honesty, impartiality, fairness and equity, and must be dedicated to the protection of the public health, safety and welfare. In the practice of their profession, engineers must perform under a standard of professional behavior which requires adherence to the highest principles of ethical conduct on behalf of the public, clients, employers and the profession.

I. Fundamental Canons

Engineers, in the fulfillment of their professional duties, shall:

1. Hold paramount the safety, health and welfare of the public in the performance of their professional duties.
2. Perform services only in areas of their competence.
3. Issue public statements only in an objective and truthful manner.
4. Act in professional matters for each employer or client as faithful agents or trustees.
5. Avoid improper solicitation of professional employment.

II. Rules of Practice

1. Engineers shall hold paramount the safety, health and welfare of the public in the performance of their professional duties.
   a. Engineers shall at all times recognize that their primary obligation is to protect the safety, health, property and welfare of the public. If their professional judgment is overruled under circumstances where the safety, health, property or welfare of the public are endangered, they shall notify their employer or client and such other authority as may be appropriate.

Continued on next page

    b. Engineers shall approve only those engineering documents which are safe for public health, property and welfare in conformity with accepted standards.

    c. Engineers shall not reveal facts, data or information obtained in a professional capacity without the prior consent of the client or employer except as authorized or required by law or this Code.

    d. Engineers shall not permit the use of their name or firm name nor associate in business ventures with any person or firm which they have reason to believe is engaging in fraudulent or dishonest business or professional practices.

    e. Engineers having knowledge of any alleged violation of this Code shall cooperate with the proper authorities in furnishing such information or assistance as may be required.

2. Engineers shall perform services only in the areas of their competence.

    a. Engineers shall undertake assignments only when qualified by education or experience in the specific technical fields involved.

    b. Engineers shall not affix their signatures to any plans or documents dealing with subject matter in which they lack competence, nor to any plan or document not prepared under their direction and control.

    c. Engineers may accept an assignment outside of their fields of competence to the extent that their services are restricted to those phases of the project in which they are qualified, and to the extent that they are satisfied that all other phases of such project will be performed by registered or otherwise qualified associates, consultants, or employees, in which case they may then sign the documents for the total project.

3. Engineers shall issue public statements only in an objective and truthful manner.

    a. Engineers shall be objective and truthful in professional reports, statements or testimony. They shall include all relevant and pertinent information in such reports, statements or testimony.

    b. Engineers may express publicly a professional opinion on technical subjects only when that opinion is founded upon adequate knowledge of the facts and competence in the subject matter.

    c. Engineers shall issue no statements, criticisms or arguments on technical matters which are inspired or paid for by interested parties, unless they have prefaced their comments by explicitly identifying the interested parties on whose behalf they are speaking, and by revealing the existence of any interest the engineer may have in the matters.

4. Engineers shall act in professional matters for each employer or client as faithful agents or trustees.

    a. Engineers shall disclose all known or potential conflicts of interest to their employers or clients by promptly informing them of any business association, interest, or other circumstances which could influence or appear to influence their judgment or the quality of their services.

    b. Engineers shall not accept compensation, financial or otherwise, from more than one party for services on the same project, or for services pertaining to the same project, unless the circumstances are fully disclosed to, and agreed to, by all interested parties.

    c. Engineers shall not solicit or accept financial or other valuable consideration, directly or indirectly, from contractors, their agents, or other parties in connection with work for employers or clients for which they are responsible.

    d. Engineers in public service as members, advisors or employees of a governmental body or department shall not participate in decisions with respect to professional services solicited or provided by them or their organizations in private or public engineering practice.

    e. Engineers shall not solicit or accept a professional contract from a governmental body on which a principal or officer of their organization serves as a member.

5. Engineers shall avoid improper solicitation of professional employment.

    a. Engineers shall not falsify or permit misrepresentation of their, or their associates', academic or professional qualifications. They shall not misrepresent or exaggerate their degree of responsibility in or for the subject matter of prior assignments. Brochures or other presentations incident to the solicitation of employment shall not misrepresent pertinent facts concerning employers, employees, associates, joint venturers or past accomplishments with the intent and purpose of enhancing their qualifications and their work.

    b. Engineers shall not offer, give, solicit or receive, either directly or indirectly, any political contribution in an amount intended to influence the award of a contract by public authority, or which may be reasonably construed by the public of having the effect or intent to influence the award of a contract. They shall not offer any gift, or other valuable consideration in order to secure work. They shall not pay a commission, percentage or brokerage fee in order to secure

Continued on next page

work except to a bona fide employee or bona fide established commercial or marketing agencies retained by them.

III. Professional Obligations

1. Engineers shall be guided in all their professional relations by the highest standards of integrity.

   a. Engineers shall admit and accept their own errors when proven wrong and refrain from distorting or altering the facts in an attempt to justify their decisions.

   b. Engineers shall advise their clients or employers when they believe a project will not be successful.

   c. Engineers shall not accept outside employment to the detriment of their regular work or interest. Before accepting any outside employment they will notify their employers.

   d. Engineers shall not attempt to attract an engineer from another employer by false or misleading pretenses.

   e. Engineers shall not actively participate in strikes, picket lines, or other collective coercive action.

   f. Engineers shall avoid any act tending to promote their own interest at the expense of the dignity and integrity of the profession.

2. Engineers shall at all times strive to serve the public interest.

   a. Engineers shall seek opportunities to be of constructive service in civic affairs and work for the advancement of the safety, health and well-being of their community.

   b. Engineers shall not complete, sign or seal plans and/or specifications that are not of a design safe to the public health and welfare and in conformity with accepted engineering standards. If the client or employer insists on such unprofessional conduct, they shall notify the proper authorities and withdraw from further service on the project.

   c. Engineers shall endeavor to extend public knowledge and appreciation of engineering and its achievements and to protect the engineering profession from misrepresentation and misunderstanding.

3. Engineers shall avoid all conduct or practice which is likely to discredit the profession or deceive the public.

   a. Engineers shall avoid the use of statements containing a material misrepresentation of fact or omitting a material fact necessary to keep statements from being misleading; statements intended or likely to create an unjustified expectation; statements containing prediction of future success; statements containing an opinion as to the quality of the engineers' services; or statements intended or likely to attract clients by the use of showmanship, puffery, or self-laudation, including the use of slogans, jingles, or sensational language or format.

   b. Consistent with the foregoing, engineers may advertise for recruitment of personnel.

   c. Consistent with the foregoing, engineers may prepare articles for the lay or technical press, but such articles shall not imply credit to the author for work performed by others.

4. Engineers shall not disclose confidential information concerning the business affairs or technical processes of any present or former client or employer without his consent.

   a. Engineers in the employ of others shall not without the consent of all interested parties enter promotional efforts or negotiations for work or make arrangements for other employment as a principal or to practice in connection with a specific project for which the engineer has gained particular and specialized knowledge.

   b. Engineers shall not, without the consent of all interested parties, participate in or represent an adversary interest in connection with a specific project or proceeding in which the engineer has gained particular specialized knowledge on behalf of a former client or employer.

5. Engineers shall not be influenced in their professional duties by conflicting interests.

   a. Engineers shall not accept financial or other considerations, including free engineering designs, from material or equipment suppliers for specifying their product.

   b. Engineers shall not accept commissions or allowances, directly or indirectly, from contractors or other parties dealing with clients or employers of the engineer in connection with work for which the engineer is responsible.

6. Engineers shall uphold the principle of appropriate and adequate compensation for those engaged in engineering work.

   a. Engineers shall not accept remuneration from either an employee or employment agency for giving employment.

   b. Engineers, when employing other engineers, shall offer a salary according to professional qualifications and the recognized standards in the particular geographical area.

   c. Engineers in sales employment shall not offer, or give engineering consultation, or designs, or advice other than specifically applying to the equipment being sold.

7. Engineers shall not compete unfairly with other engineers by attempting to obtain employment or advancement or professional engagements by taking advantage of a salaried position,

Continued on next page

by criticizing other engineers, or by other improper or questionable methods.

    a. Engineers shall not request, propose, or accept a professional commission on a contingent basis under circumstances in which their professional judgment may be compromised.

    b. Engineers in salaried positions shall accept part-time engineering work only at salaries not less than that recognized as standard in the area.

    c. Engineers shall not use equipment, supplies, laboratory, or office facilities of an employer to carry on outside private practice without consent.

8. Engineers shall not attempt to injure, maliciously or falsely, directly or indirectly, the professional reputation, prospects, practice or employment of other engineers, nor indiscriminately criticize other engineers' work. Engineers who believe other engineers are guilty of unethical or illegal practice shall present such information to the proper authority for action.

    a. Engineers in private practice shall not review the work of another engineer for the same client, except with the knowledge of such engineer, or unless the connection of such engineer with the work has been terminated.

    b. Engineers in governmental, industrial or educational employ are entitled to review and evaluate the work of other engineers when so required by their employment duties.

    c. Engineers in sales or industrial employ are entitled to make engineering comparisons of represented products with products of other suppliers.

9. Engineers shall accept personal responsibility for all professional activities.

    a. Engineers shall conform with state registration laws in the practice of engineering.

    b. Engineers shall not use association with a nonengineer, a corporation, or partnership, as a "cloak" for unethical acts, but must accept personal responsibility for all professional acts.

10. Engineers shall give credit for engineering work to those to whom credit is due, and will recognize the proprietary interests of others.

    a. Engineers shall, whenever possible, name the person or persons who may be individually responsible for designs, inventions, writings, or other accomplishments.

    b. Engineers using designs supplied by a client shall recognize that the designs remain the property of the client and may not be duplicated by the engineer for others without express permission.

c. Engineers, before undertaking work for others in connection with which the engineer may make improvements, plans, designs, inventions, or other records which may justify copyrights or patents, should enter into a positive agreement regarding ownership.

d. Engineers' designs, data, records, and notes referring exclusively to an employer's work are the employer's property.

11. Engineers shall cooperate in extending the effectiveness of the profession by interchanging information and experience with other engineers and students, and will endeavor to provide opportunity for the professional development and advancement of engineers under their supervision.

a. Engineers shall encourage engineering employees' efforts to improve their education.

b. Engineers shall encourage engineering employees to attend and present papers at professional and technical society meetings.

c. Engineers shall urge engineering employees to become registered at the earliest possible date.

d. Engineers shall assign a professional engineer duties of a nature to utilize full training and experience, insofar as possible, and delegate lesser functions to subprofessionals or to technicians.

e. Engineers shall provide a prospective engineering employee with complete information on working conditions and proposed status of employment, and after employment will keep employees informed of any changes.

"By order of the United States District Court for the District of Columbia, former Section 11(c) of the NSPE Code of Ethics prohibiting competitive bidding, and all policy statement opinions, rulings or other guidelines interpreting its scope, have been rescinded as unlawfully interfering with the legal right of engineers, protected under the antitrust laws, to provide price information to prospective clients; accordingly, nothing contained in the NSPE Code of Ethics, policy statements, opinions, rulings or other guidelines prohibits the submission of price quotations or competitive bids for engineering services at any time or in any amount."

Statement by NSPE Executive Committee

In order to correct misunderstandings which have been indicated in some instances since the issuance of the Supreme Court decision and the entry of the Final Judgment, it is noted that in its decision of

Continued on next page

April 25, 1978, the Supreme Court of the United States declared: "The Sherman Act does not require competitive bidding."

It is further noted that as made clear in the Supreme Court decision:

1. Engineers and firms may individually refuse to bid for engineering services.
2. Clients are not required to seek bids for engineering services.
3. Federal, state and local laws governing procedures to procure engineering services are not affected, and remain in full force and effect.
4. State societies and local chapters are free to actively and aggressively seek legislation for professional selection and negotiation procedures by public agencies.
5. State registration board rules of professional conduct, including rules prohibiting competitive bidding for engineering services, are not affected and remain in full force and effect. State registration boards with authority to adopt rules of professional conduct may adopt rules governing procedures to obtain engineering services.
6. As noted by the Supreme Court, "nothing in the judgment prevents NSPE and its members from attempting to influence governmental action. . . ."

Note: In regard to the question of application of the Code to corporations vis-à-vis real persons, business form or type should not negate nor influence conformance of individuals to the Code. The Code deals with professional services, which services must be performed by real persons. Real persons in turn establish and implement policies within business structures. The Code is clearly written to apply to the engineer and it is incumbent on a member of NSPE to endeavor to live up to its provisions. This applies to all pertinent sections of the Code.

*Reprinted courtesy of the National Society of Professional Engineers, NSPE Publication No. 1102 as revised, January 1981.

PROBLEMS

4.1 What procedures can consulting engineers use to advertise their practice? Review the NSPE Code of Ethics presented at the end of this chapter before answering this question.

4.2 It is often argued that engineers in large industries do not have direct contact with the public. Present an argument for the registration of *all* engineers.

4.3 A college senior is interviewing with several companies for an engineering position. After interviewing with company B, the student makes a verbal job commitment in response to a tentative job offer from one of the industrial representatives of company B. The following week, company C sends the student an airline round-trip ticket for an upcoming plant trip. Since the student has never visited the region where company C is located, and since the job offer from company B has not yet been formalized, is it ethical for the student to take this trip? Discuss.

4.4 Is it permissible for engineers to bid against other engineers for contracts to do engineering work? Supreme Court decisions of a recent nature have had a bearing on this question.

4.5 Interview a professional engineer and get his or her views on engineering professionalism.

4.6 One of the more interesting cases in ethical studies concerns the DC-10 cargo door. Do a thorough investigation into this case, delving into the ramifications of the original design and the later design modifications. This case is discussed in some detail in the March and April 1974 issues of *Aviation Week and Space Technology* and in the June and July 1975 issues of the same periodical.

4.7 In your opinion, why must young engineers have at least four years of professional experience before they can apply for registration as a professional engineer? Doesn't four or five years of engineering education provide the necessary experience?

4.8 Interview medical doctors or attorneys at law and request their opinion of engineering as a profession. Also ask them to distinguish the professional differences between their profession and engineering.

PART THREE
Fundamental Engineering Skills

CHAPTER 5
THE HAND-HELD CALCULATOR

CHAPTER 6
DIMENSIONS AND UNITS

CHAPTER 7
THE TECHNICAL LIBRARY

5

The Hand-Held Calculator

CONTENTS

Introduction
Questions about Calculators
Basic Features of a Calculator
Essential Calculator Functions
Nonessential but Useful Calculator Functions
Algebraic Operating System (AOS)
Reversed Polish Notation (RPN)
Some Common Errors

> With the advent of the hand-held calculator, much of the computational drudgery previously required of students in problem solving has been eliminated. Yet there is danger if we take this marvelous device for granted. Bad calculator habits are easily developed, and mistakes can be made that are not obvious in the sequence of calculator operations. Students must be aware of these calculation subtleties.

INTRODUCTION

Engineers and engineering and technology students alike count hand-held calculators as one of the significant blessings of modern technology. Calculations that previously were tedious and time consuming have become almost routine. Programmability, now available on even the less expensive calculators, has eased the burden of repetitive computations even further. Many colleges of engineering now require their students to have programmable calculators. Some even require their students to purchase microcomputers.

Because calculator proficiency is an essential problem-solving prerequisite, this chapter examines the basic operational rules that govern calculator manipulations. Students experienced with hand-held calculators may find that a detailed study of this chapter is unnecessary. A quick review of the subject matter and practice with the problems at the end of this chapter may prove sufficient.

QUESTIONS ABOUT CALCULATORS

The following are questions often raised about scientific calculators. We address them somewhat superficially in this section. Some of these questions cannot be answered in a universal way because of the different needs and preferences of individuals. You may receive more satisfactory answers as you read and ponder the following sections, which discuss calculator operations and capabilities in some depth.

WHICH SCIENTIFIC CALCULATOR SHOULD I PURCHASE? *The best one you can afford!* Often, your economic situation will guide this choice. But price should not be the only reason for your choice. It is better to stretch your

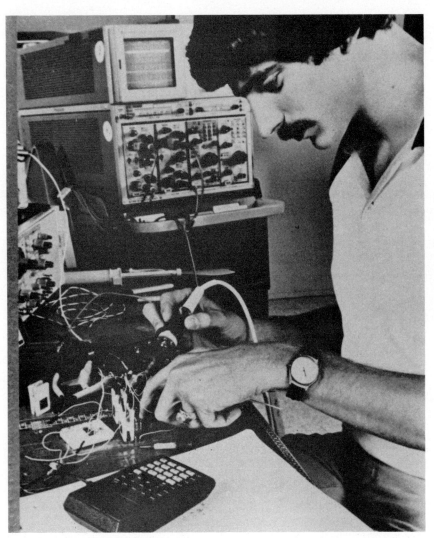

Calculators, indispensable to the engineer, technologist, and technician, can be found wherever engineering activities are taking place. (Courtesy of Data General)

budget a little than to buy a calculator that will burden you for semesters to come. Fortunately, there are inexpensive scientific calculators that meet the minimum requirements for calculator functions. Certainly, your decision will be governed by several factors: cost, type of logic, programmability, whether you are absolutely certain about an engineering career, and so on.

IS THE CALCULATOR I HAVE SATISFACTORY? This question will be addressed more completely in the following sections; at least, you will be able to make the comparisons necessary to answer this question for yourself. We will say that if you have a scientific calculator made by Casio, Hewlett-

An advanced reverse Polish notation (RPN) calculator, the HP-41C made by Hewlett-Packard is a powerful programmable calculator supported by several peripheral devices. (Courtesy of Hewlett-Packard)

Packard (HP), Radio Shack, Sharp, or Texas Instruments (TI)—in other words, a major manufacturer of scientific hand-held calculators—it is probably satisfactory.

SHOULD MY CALCULATOR BE PROGRAMMABLE? Programmability may not be essential to you in the first or second year of your undergraduate education. Frankly, many technical students graduate without ever owning or knowing how to use a programmable calculator. Nevertheless, in their junior and senior years, students will be assigned many problems of an iterative nature for which programmability would be a real plus. Many students familiar with their programmable calculator will also be able to write short programs to verify their computer output from some FORTRAN or BASIC program that they have written. Current trends indicate that colleges of engineering are expecting—even requiring—their students to have programmable calculators. Thus, in our opinion, every engineering student should have one. Cost is no longer a limiting factor, since programmable calculators can be bought for less than $100, and some for even less than $50.

The Sharp EL-5100 calculator displays equations alphanumerically. (Courtesy of Sharp)

IS REVERSE POLISH NOTATION (RPN) BETTER CALCULATOR LOGIC THAN THAT OF THE ALGEBRAIC SYSTEM? This is a difficult question to answer; both logic systems have strong proponents and detractors. On the average, fewer keystrokes will be required to perform routine mathematical operations on RPN calculators like Hewlett-Packard (HP) than on algebraic-system calculators like Texas Instruments (TI), and intermediate answers always have significance. On the other hand, proponents of algebraic-system calculators argue that "algebraic" operations more closely resemble the order in which we have been taught to perform algebraic calculations. In addition, they argue that the algebraic system with hierarchy (which most algebraic calculators have and which we will refer to as the algebraic operating system, or AOS) strongly relates to the operations that students will learn in computer programming. There are a number of other arguments; some will be discussed later. The only conclusion that we can reach is that students will have to make their own decisions based on their personal preferences.

BASIC FEATURES OF A CALCULATOR

What should be the essentials for a scientific hand-held calculator? Certainly this is a debatable question, one that we are somewhat reluctant to answer. Nevertheless, we feel that it is important that you have something to compare your calculator with. Before we present our opinion of what these essentials should be, we define several terms that may not be familiar to you; also note the calculator elements identified in Figure 5.1.

KEY This protrusion (or segregated area on a touch-sensitive calculator) is pressed to activate a calculator circuit. The *primary* or *first function* is shown on the top face of the key; the *secondary* or *second function* is indicated on the face of the calculator next to the key or on a side face of the key. To use

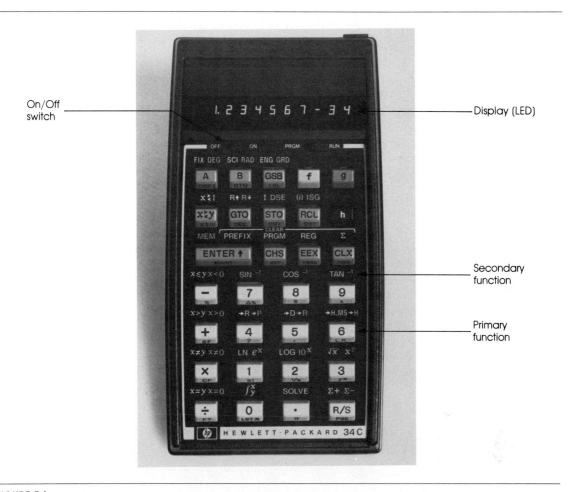

FIGURE 5.1
Fundamental elements of a calculator. (Courtesy of Hewlett-Packard)

the second function of a key it is necessary first to press an additional key identified by color and/or by alphanumerics like 2nd F, 2nd, f, g, and so forth.

LED/LCD   There are two types of displays, LED and LCD. **LED** stands for light-emitting diode (usually red); **LCD** stands for liquid crystal display. LEDs are difficult to see in normal outdoor light but easy to see in darker areas; LCDs normally display black alphanumerics against a gray or yellow background and are easy to see in areas that are reasonably well lit (Figure 5.2).

MAGNETIC CARD   The magnetic card is a magnetic strip that is inserted into the calculator, either to store programmed instructions for later use or to have stored programmed instructions read into the calculator. Only some of

FIGURE 5.2
An LCD display. Some calculator manufacturers offer calculators having both LED and LCD displays. (Courtesy of Hewlett-Packard)

LCD display

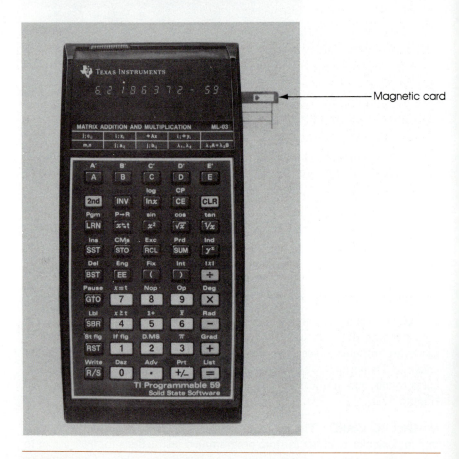

Magnetic card

FIGURE 5.3
Calculator with magnetic card capability. (Courtesy of Texas Instruments)

Program libraries are often available for programmable calculators. (Courtesy of Texas Instruments)

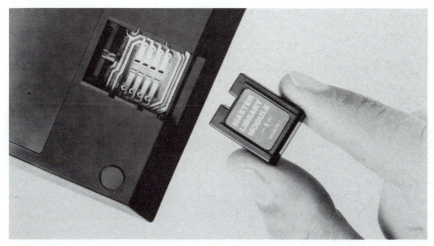

FIGURE 5.4
A calculator module. (Courtesy of Texas Instruments)

the more advanced calculators have this capability (Figure 5.3). The **card reader** is that part of the calculator that reads or programs the magnetic card.

CALCULATOR MODULES A calculator module is a small memory component that plugs into the calculator (Figure 5.4). **Application** or **library modules** (available in some HP calculators) increase the number of data storage registers or lines of program memory accessible to the calculator user for programming or number storage.

MEMORY OR STORAGE REGISTER The **memory** is a temporary storage area for numbers or programming instructions.

PROGRAM/SOFTWARE A **program** is a set of instructions that orders the operations performed by the calculator. A program is often called the **software,** particularly when it is widely used. Note that function keys like $\boxed{\text{sin}}$, $\boxed{\text{cos}}$, and $\boxed{e^x}$ are themselves programs built into the calculator.

ESSENTIAL CALCULATOR FUNCTIONS

In Table 5.1 we present what we consider to be the essential keys of a scientific calculator. In all likelihood, if your calculator is programmable, it will have these capabilities. If your calculator is not programmable, you may wish to test its capabilities against this checklist. Note that in this chapter we do *not* distinguish between the first and second function keys by different key background colors (usually black and white) as do many instruction manuals. For example, in the instruction manual for a TI-55 II calculator, the e^x key is denoted by $\boxed{e^x}$, whereas the second function x^2 key is denoted by $\boxed{x^2}$. Each student will have to determine which functions are first and which are second for his or her calculator and then press the keys appropriately.

TABLE 5.1   Essential Calculator Keys

Function	Symbol	Explanation
Numbers	$\boxed{0}$–$\boxed{9}$	Basic digit keys.
Decimal point	$\boxed{\cdot}$	Unless pressed, the decimal is "right justified"; in other words, it is always to the right of the last digit keyed in.
Change sign	$\boxed{+/-}$ or $\boxed{\text{CHS}}$	Reverses the sign of the last number entered.
Addition; subtraction	$\boxed{+}$; $\boxed{-}$	Basic keys.
Multiplication; division	$\boxed{\times}$; $\boxed{\div}$	Basic keys.
Enter number	$\boxed{\text{ENTER} \uparrow}$	Required on RPN calculators to enter a number into one of the "stack" registers above the display register; not required on AOS calculators.
Equals sign	$\boxed{=}$	Completes all pending operations on AOS calculators; not required on RPN calculators.
Power of 10 notation	$\boxed{\text{EE}}$ or $\boxed{\text{EEX}}$ or $\boxed{\text{EXP}}$	Allows you to enter a number times an integer power of 10. Subsequent operations will display numbers in scientific notation, in other words, as a

TABLE 5.1 Essential Calculator Keys (continued)

Function	Symbol	Explanation
		number between 1 and 10 times the appropriate integer power of 10. In some calculators, you can convert back to standard notation by pressing the [INV] and [EE] keys.
Pi	π	Recalls the value of pi (3.141 592 6 . . .).
Trigonometric functions	[SIN]; [COS]; [TAN]	Determines the trigonometric value of the argument (an angle) keyed into the calculator; the argument can be in either positive or negative degrees or radians, and the trigonometric value can be either positive or negative, depending on which quadrant the angle lies within. The reciprocal functions secant, cosecant, and cotangent are determined by using the reciprocal key [1/x]. The trigonometric keys should determine the values of arguments, both positive and negative, and *greater than* $\lvert\pm360°\rvert$ or $\lvert\pm2\pi\rvert$ radians.
Inverse trigonometric functions	[SIN^{-1}] or [INV] [SIN]; [COS^{-1}] or [INV] [COS]; [TAN^{-1}] or [INV] [TAN]	Arcsine, sin^{-1}, or inverse sine all mean to find the angle whose sine has some value (similarly stated for cos^{-1} or tan^{-1}). Keys must give correct values for arguments that are both positive and negative.
Reciprocal key	[1/x]	Determines the reciprocal of x; for example, given $x = 2$, $1/x$ gives 0.5. The reciprocal key is often used to find the reciprocal trigonometric functions; for example, to find the cosecant of 30°, press [3]

TABLE 5.1 Essential Calculator Keys (continued)

Function	Symbol	Explanation
Square, square root	x^2; $\sqrt{x}$ or $\sqrt{}$	$\boxed{0}$ $\boxed{\text{SIN}}$ and then press $\boxed{1/x}$ to get 2. Even though you can determine either the square or square root using the $\boxed{y^x}$ key, the calculator should have separate square and square root keys, since both are frequently evaluated.
Exponent	y^x	Calculates the value of some positive base number y having an exponent x. Exponent can be positive or negative and have values greater than or less than 1. To find $\sqrt[x]{y}$, use either the $\boxed{\sqrt[x]{y}}$ key, the $\boxed{y^x}$ key in combination with the $\boxed{1/x}$ key, or the $\boxed{\text{INV}}$ $\boxed{y^x}$ keys, depending on the calculator.
Logarithms	$\boxed{\text{LOG}}$; $\boxed{\text{LN}}$ or $\boxed{\text{LNX}}$	$\boxed{\text{LOG}}$ determines the logarithm to the base 10 of numbers greater than zero. $\boxed{\text{LN}}$ or $\boxed{\text{LNX}}$ determines the logarithm of numbers greater than zero to the base e (e = 2.718 28 . . .).
Base 10 exponent	10^x or $\boxed{\text{INV}}$ $\boxed{\text{LOG}}$	Although you could determine 10^x using the $\boxed{y^x}$ key, 10^x is a common mathematical operation and should have a separate key.
Base e exponent	e^x or $\boxed{\text{INV}}$ $\boxed{\text{LN}}$ or $\boxed{\text{INV}}$ $\boxed{\text{LNX}}$	Exponents to the base e are frequently evaluated and require a separate key.
Clear	$\boxed{\text{CLR}}$ or $\boxed{\text{C}}$ or $\boxed{\text{CLX}}$	Clears display on both RPN and algebraic-system calculators and all pending operations on AOS calculators.
Clear entry	$\boxed{\text{CE}}$	On algebraic-system

TABLE 5.1 Essential Calculator Keys (continued)

Function	Symbol	Explanation
		calculators, the $\boxed{\text{CE}}$ key clears the last entry if a number. On some calculators it clears a mathematical operation if it is the last entry.
Memory register(s)	$\boxed{\text{STO}}$ or $\boxed{\text{X→M}}$; $\boxed{\text{RCL}}$ or $\boxed{\text{RM}}$	$\boxed{\text{STO}}$ followed by a number (if more than one memory register is available) is the keystroke sequence to store a number in a memory register. $\boxed{\text{RCL}}$ followed by a number displays the stored number. A calculator must have *at least one* storage register.
Last x	$\boxed{\text{LST } x}$ or $\boxed{\text{LAST } x}$	Every time a function key is pressed, the last number in the display before the function key is pressed goes into the LAST x register. LAST x is available on Hewlett-Packard RPN calculators.

In addition to the features listed in Table 5.1, your calculator should also have the following capabilities:

1. The calculator should be able to manipulate positive and negative large and small numbers, preferably numbers X in the range $1 \times 10^{-99} < |X| < 10 \times 10^{99}$.
2. The calculator should be able to perform the conversion from polar coordinates (r, θ) to rectangular coordinates (x, y) and vice versa.
3. The calculator should be able to calculate the trigonometric values of angles in either degrees or radians. Some scientific calculators also allow angles to be in **grads** (400 grads per revolution), although this is not essential.
4. Algebraic-system calculators should have algebraic **hierarchy;** this will be discussed in one of the following sections.

NONESSENTIAL BUT USEFUL CALCULATOR FUNCTIONS

Additional keys that often prove useful to engineers are described in Table 5.2. They are not considered essential because their functions can be implemented through the essential keys already available and because they are required only occasionally. However, many of these keys can be found on even the less expensive calculators.

TABLE 5.2 Nonessential Calculator Keys

Function	Symbol	Explanation
Parentheses	$\boxed{(}$; $\boxed{)}$	Builds in an operational hierarchy based on the "nesting" of parentheses. The mathematical operations within the innermost parentheses will be performed first. Parentheses keys will be found only on algebraic-system calculators, not on RPN calculators.
Statistical keys	$\boxed{\Sigma}$ or $\boxed{\Sigma+}$ or $\boxed{\Sigma x}$; $\boxed{\bar{x}}$; $\boxed{s}$ or $\boxed{\sigma_n}$ or $\boxed{\sigma_{n-1}}$; $\boxed{\sigma^2}$	These statistical keys compute the sum total of a set of numbers by the key $\boxed{\Sigma}$, their mean value $\boxed{\bar{x}}$, and their standard deviation $\boxed{s}$ or variation $\boxed{\sigma^2}$, depending on the calculator.
Factorial key	$\boxed{n!}$ or $\boxed{x!}$	A useful probability function key that calculates the factorial of an integer n or x; for example, $3! = 1 \cdot 2 \cdot 3 = 6$. Zero factorial is defined to be 1, $0! = 1$.
Engineering notation	$\boxed{\text{ENG}}$ or $\boxed{\text{Eng}}$	Places any positive (negative) number into a number between $1(-1)$ and $1\ 000\ (-1\ 000)$ times 10 raised to a power that is a positive or negative integral multiple of 3. For example, $-0.056\ 7$ in engineering notation is -56.7×10^{-3}.
Fix number of significant digits	$\boxed{\text{FIX}}$ or $\boxed{\text{Fix}}$	Fixes the number of significant digits that will be displayed on the calculator. Internal numbers will be automatically rounded off for the display. Note that when HP calculators are switched on, the display automatically rounds off numbers to two decimal places. Internally, however, the original

TABLE 5.2 Nonessential Calculator Keys (continued)

Function	Symbol	Explanation
		value is retained. The $\boxed{\text{FIX}}$ key must be used if more than two decimal places are to be displayed.
Memory sum; memory exchange	$\boxed{\text{SUM}}$ or $\boxed{\text{M+}}$; $\boxed{\text{EXC}}$	$\boxed{\text{SUM}}$ or $\boxed{\text{M+}}$ allows you to add the displayed number directly to the number in memory. $\boxed{\text{EXC}}$ allows you to exchange the displayed number with the number in memory.
Exchange key	$\boxed{x \gtrless y}$	On HP calculators, pressing the $\boxed{x \gtrless y}$ key exchanges the contents of the display register (X register) with the next register in the stack (Y register).

ALGEBRAIC OPERATING SYSTEM (AOS)

The algebraic system used in Texas Instruments calculators is a hierarchical system; in other words, there is a preferred order of mathematical operations, called the algebraic operating system (AOS). Because most other algebraic-system calculators utilize similar hierarchical logic, we will confine our discussion to AOS.

AOS without Parentheses

In the absence of parentheses keys, AOS operations are performed in the following order:

1. **Single-variable keys** act immediately on the number displayed. Examples of single-variable keys are $\boxed{1/x}$, $\boxed{e^x}$, $\boxed{\log}$, $\boxed{\ln x}$, $\boxed{\sin}$, $\boxed{\cos}$, $\boxed{\tan}$, $\boxed{x}$, $\boxed{x^2}$, and $\boxed{n!}$.
2. **Exponentiation** is performed as soon as single-variable functions are completed. Examples are $\boxed{y^x}$ and the root keys $\boxed{y^x}$, $\boxed{\sqrt{x}}$, $\boxed{\text{INV}}$ $\boxed{y^x}$, or $\boxed{\sqrt[x]{y}}$.
3. **Multiplications** and **divisions** are completed next.
4. **Additions** and **subtractions** are completed next.
5. The **equals** key $\boxed{=}$ completes all pending operations.

To demonstrate AOS hierarchy, we will work several examples for you. You can verify whether your calculator has AOS logic by following the keystroke sequence indicated and then comparing your answer with the one shown.

Examples

EXAMPLE 5.1: $2 + 3 \times 1.2 = ?$

Keystrokes: $\boxed{2}\,\boxed{+}\,\boxed{3}\,\boxed{\times}\,\boxed{1}\,\boxed{\bullet}\,\boxed{2}\,\boxed{=}$ $\underline{\quad 5.6 \quad}$
(Display)

EXAMPLE 5.2: $3.67 \times 25^{3.2} \times \tan 126° = ?$

Keystrokes: $\boxed{3}\,\boxed{\bullet}\,\boxed{6}\,\boxed{7}\,\boxed{\times}\,\boxed{2}\,\boxed{5}\,\boxed{y^x}\,\boxed{3}\,\boxed{\bullet}\,\boxed{2}\,\boxed{\times}\,\boxed{1}\,\boxed{2}$
$\boxed{6}\,\boxed{\tan}\,\boxed{=}$ $\underline{-150.2495}$
(Display)

EXAMPLE 5.3: $2^{\sin 30°} = ?$

Keystrokes: $\boxed{2}\,\boxed{y^x}\,\boxed{3}\,\boxed{0}\,\boxed{\sin}\,\boxed{=}$ $\underline{1.4142135}$
(Display)

AOS with Parentheses $\boxed{(}$, $\boxed{)}$

Parentheses allow you to specify the order in which mathematical expressions will be evaluated, namely, from the innermost parentheses in which the operations are evaluated first to the outermost parentheses which are performed last. Within parentheses, the normal AOS hierarchy will prevail. The advantage of parentheses is that they allow the user to cluster operations and to obtain intermediate answers. The disadvantage is that the user may lean too much on parentheses and not use the hierarchy already built into the calculator. Calculator users who fall into this habit use more keystrokes than necessary.

There are two important rules governing the use of parentheses:

- As soon as any set of parentheses is closed, all the calculations inside are performed immediately.
- Pressing the $\boxed{=}$ key will complete all pending operations up to that point. Press this only when you want the final answer.

We demonstrate the advantages and disadvantages of parentheses in the following examples by comparing the kind and number of keystrokes required with and without parentheses.

Examples

EXAMPLE 5.4 $(3 + 1) \times (4 - 3) = ?$

Keystrokes with Parentheses:

$$\boxed{(}\;\boxed{3}\;\boxed{+}\;\boxed{1}\;\boxed{)}\;\boxed{\times}\;\boxed{(}\;\boxed{4}\;\boxed{-}\;\boxed{3}\;\boxed{)}\;\boxed{=}\quad \underline{\quad 4.\quad}$$
(Display)

Keystrokes without Parentheses:

$$\boxed{3}\;\boxed{+}\;\boxed{1}\;\boxed{=}\;\boxed{STO}\;\boxed{4}\;\boxed{-}\;\boxed{3}\;\boxed{=}\;\boxed{\times}\;\boxed{RCL}\;\boxed{=}\quad \underline{\quad 4.\quad}$$
(Display)

Comment: Each keystroke sequence requires twelve keystrokes; yet, with parentheses you do not have to remember that a number is stored in memory.

EXAMPLE 5.5 $\dfrac{(8 + 4) + (9 \times 19)}{(3 + 10 \div 7) \times 2} = ?$

Keystrokes with Parentheses:

$$\boxed{(}\;\boxed{(}\;\boxed{8}\;\boxed{+}\;\boxed{4}\;\boxed{)}\;\boxed{+}\;\boxed{(}\;\boxed{9}\;\boxed{\times}\;\boxed{1}\;\boxed{9}\;\boxed{)}\;\boxed{)}$$
$$\boxed{\div}\;\boxed{(}\;\boxed{(}\;\boxed{3}\;\boxed{+}\;\boxed{1}\;\boxed{0}\;\boxed{\div}\;\boxed{7}\;\boxed{)}\;\boxed{\times}\;\boxed{2}\;\boxed{)}\;\boxed{=}\quad 231.67741$$
(Display)

Keystrokes without Parentheses:

$$\boxed{3}\;\boxed{+}\;\boxed{1}\;\boxed{0}\;\boxed{\div}\;\boxed{7}\;\boxed{=}\;\boxed{\times}\;\boxed{2}\;\boxed{=}\;\boxed{STO}\;\boxed{8}\;\boxed{+}\;\boxed{4}\;\boxed{=}$$
$$\boxed{+}\;\boxed{9}\;\boxed{\times}\;\boxed{1}\;\boxed{9}\;\boxed{=}\;\boxed{\div}\;\boxed{RCL}\;\boxed{=}\quad 231.67741$$
(Display)

Comments: In this example twenty-seven keystrokes are required with parentheses and twenty-two keystrokes are required without parentheses. With parentheses you can separate the numerator calculations from the denominator calculations. Without parentheses, more strategy is required in determining where to start the calculations. In addition, frequent use of the $\boxed{=}$ key is required to complete pending operations as necessary.

To understand some of the pitfalls that might occur in calculations of this type, answer the following questions for Example 5.5.

With Parentheses:

- What happens if you omit the outermost parentheses in either the denominator or the numerator?

Without Parentheses:

- Why should you begin the calculation with the denominator? This question is especially important if your calculator does not have a memory exchange key $\boxed{\text{EXC}}$.
- What happens to your answer if you omit any of the intermediate $\boxed{=}$ keystrokes?

REVERSED POLISH NOTATION (RPN)

Jan Lukasiewicz, born December 21, 1878, in the Polish city of Lwow, is often referred to as the "father of RPN." A prolific mathematical scholar, Lukasiewicz developed what is now known as **Polish notation.** This notation simplifies the evaluation of arithmetic expressions by eliminating parentheses. As implemented in HP calculators, the operational sequence is reversed; thus the name **reverse Polish notation.** The mathematical operation is specified after the numbers instead of before; for example, $\boxed{+, 2, 3 = 5}$ would be the Polish notation sequence of operations and $\boxed{2, 3, + \ = 5}$ would be the sequence as implemented in reverse Polish notation.

Hewlett Packard implements RPN through a **four-register stack** and the $\boxed{\text{ENTER} \uparrow}$ key. Each number entered simply moves the last number in the display up the stack ladder to the next register. **Automatic hierarchy** is contained in this stack ladder by virtue of relative position.

We illustrate the basic principles of RPN by following the movement of numbers through the stack in Figure 5.5 for the simple addition problem $3 + 4 = 7$. The display register is the bottom, or X, register. The other three registers, the Y, Z, and T registers, follow in ascending order. Note that once the number 3 is keyed in, it is displayed in the X register. It then must be entered by the key $\boxed{\text{ENTER} \uparrow}$ to move it to the next register, the Y register. Until another number is keyed in, the 3 remains in both the X and Y registers. When 4 is keyed in, the 3 in the display register is replaced by 4. Pressing the $\boxed{+}$ key automatically performs the addition of 3 in the Y register to 4 in the X register, leaving the Y register empty and displaying the result of the mathematical operation, 7.

In summary, the enter key $\boxed{\text{ENTER} \uparrow}$ moves numbers up the stack ladder while function keys like $\boxed{\times}$, $\boxed{\div}$, $\boxed{+}$, and so forth, move numbers down the ladder. Single-variable keys like $\boxed{x^2}$, $\boxed{\text{SIN}}$, and so forth, operate on the

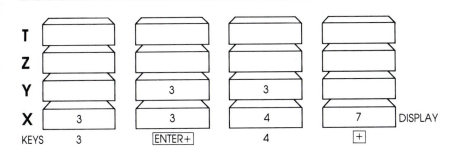

FIGURE 5.5

RPN implementation of 3 + 4 = 7

number in the display register, replacing it by the result of the mathematical operation. Keys like the $\boxed{y^x}$, $\boxed{+}$, and $\boxed{\times}$ keys operate on the numbers in the X and Y registers, displaying the result in the X register.

A slightly more difficult problem involves the calculation

$$\left[1 + 0.2\left(\frac{350}{661.5}\right)^2\right]^{3.5}$$

The sequence of RPN keystrokes is shown in Figure 5.6, along with the stack contents after each keystroke. With RPN logic it is usually recommended that calculation problems be attacked from the innermost parentheses first. This is reflected in the keystroke sequence of Figure 5.6. Note that after a functional key is pressed, one should not use the $\boxed{\text{ENTER} \uparrow}$ key to raise numbers in the stack—they are automatically raised when a new number is keyed in.

The $\boxed{\text{LAST } x}$ key allows the user to recover the last number in the display register after a function key has been pressed. This proves very useful if an error has been made. To illustrate, we return to the first problem we considered, 3 + 4 = 7. Let us suppose that we actually intended to add 5 to 3 instead of 4. Following the register contents in Figure 5.7, we see that the 4 is placed in the LAST x register after the $\boxed{+}$ key is pressed. From there it is recovered and then subtracted from 7 to recover the number 3. The 5 is then added to 3 to get the correct sum, which is 8.

Two other keys, the $\boxed{x \geq y}$ key and the $\boxed{R \downarrow}$ key, offer means to rearrange arguments in the stack. The $\boxed{x \geq y}$ key exchanges the contents in the X register with the contents in the Y register. The *rolldown key,* as the $\boxed{R \downarrow}$ key is usually called, simply moves the contents of each register down to the next register except for the contents of the display, or X register, which is moved to the top register, the T register.

These three particular keys prove their worth in more complex operations, where the ability to review register contents, reverse operations, and rearrange the stack often saves the user from having to restart a calculation sequence once an error has been made.

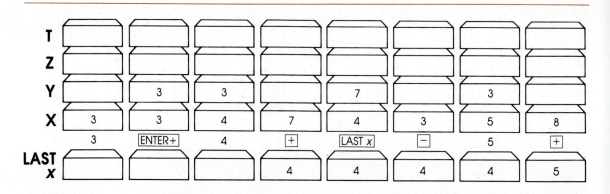

FIGURE 5.6
RPN keystrokes for $[1 + 0.2 (350/661.5)^2]^{3.5}$

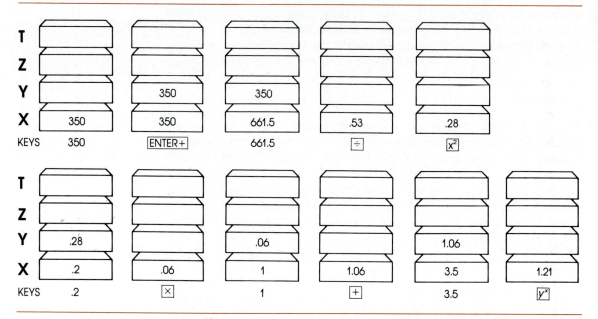

FIGURE 5.7
Correcting an error by the $\boxed{\text{LAST } x}$ key.

SOME COMMON ERRORS

Even the more experienced calculator users make some of the common errors we will discuss here. Although these errors often result from unfamiliarity with the calculator, they also can be caused by mental lapses. The result is numbers that may bear no resemblance to what they should be and engineering answers that often violate common sense. The most common errors involve the *angle mode, hierarchy mistakes, rounding off prematurely,* and *mixing arguments.*

ANGLE MODE   Probably the most common mistake made by all users of scientific calculators is performing trigonometric calculations with the calculator in the wrong angle mode. For example, if the calculator is in the radian mode and you wish to find sin 30°, pressing $\boxed{3}\,\boxed{0}\,\boxed{\text{SIN}}$ will display -0.98803, clearly not the desired answer of 0.5. Fortunately, many new calculators now display the current angle mode so that this common mistake can be avoided. If your calculator doesn't have this feature, then you must be especially alert to the possibility of this error.

HIERARCHY MISTAKES   Errors in mathematical operations can occur when we forget the hierarchy rules for our calculator. Consider the calculation $(3 + 4) \times 5 = 35$ as performed by an AOS calculator. Forgetting that multiplication/division are of higher priority than addition/subtraction would lead to calculation error if the following keystrokes were used:

$$\boxed{3}\,\boxed{+}\,\boxed{4}\,\boxed{\times}\,\boxed{5}\,\boxed{=} \quad \frac{23}{\text{(Display)}}$$

To perform the keystroke sequence correctly, it is necessary to press the $\boxed{=}$ key *after* the 4 is entered and then multiply by 5. But be warned: Pressing the $\boxed{=}$ key prematurely can also lead to hierarchy mistakes, particularly when parentheses keys are being used, because the $\boxed{=}$ key completes all pending operations.

ROUNDING OFF PREMATURELY   Unless you are manipulating numbers to a fixed number of significant digits, it is best to leave the numbers in the calculator rather than writing them down as displayed for subsequent use. For this reason it is necessary for a scientific calculator to have at least one memory—preferably more. Rounding off can cause significant errors when calculations involve the differences between numbers of almost the same size.

MIXING ARGUMENTS   The $\boxed{y^x}$ key can give erroneous results if the base number y is entered incorrectly as the exponent x or vice versa. In AOS calculators, the first number entered is the exponent x. Similarly, in RPN calculators the first number is shifted up to the Y register and is the base number y, whereas the second number displayed in the X register is the exponent x. Other errors can occur in the use of the polar-to-rectangular conversion keys when the wrong sequence of numbers is entered.

PRACTICAL EXAMPLES

Multiplication

1. $(100.7)(0.021) = 2.11$
2. $(\pi)(315.2) = 9.90 \times 10^2$
3. $(1.2 \times 10^3)(0.56) = 6.72 \times 10^2$
4. $(47.2)(67\ 203) = 3.17 \times 10^6$
5. $(2.8 \times 10^{-4})(28 \times 10^5) = 7.84 \times 10^2$
6. $(183.2)(2.2)(0.031) = 1.25 \times 10^1$
7. $(0.005\ 26)(0.356\ 1) = 1.87 \times 10^{-3}$
8. $(85.3)(81.2)(0.081) = 5.61 \times 10^2$

Division

1. $(85.67) \div (0.025) = 3.43 \times 10^3$

2. $(1.29 \times 10^{25}) \div (9.12 \times 10^{15}) = 1.41 \times 10^9$

3. $(0.085) \div (1.26 \times 10^{-4}) = 6.75 \times 10^2$

4. $\dfrac{\pi}{(1.56 \times 10^2)(0.73)} = 2.75 \times 10^{-2}$

5. $(6.27)^{-1}(16.32) = 2.50$

6. $(18.2 \times 10^5)(\pi)^{-1}(6.02 \times 10^{-3}) = 3.49 \times 10^3$

7. $\dfrac{86.32}{(1.03 \times 10^{-2})(121.2)} = 5.91 \times 10^1$

8. $(123.6)^{-1}(0.056\ 67) = 4.58 \times 10^{-4}$

Multiplication and Division

1. $\dfrac{(\pi)(106.2)(60.7)}{(123.9)} = 1.63 \times 10^2$

2. $\dfrac{(0.005\ 56)(0.111)}{(1.25 \times 10^5)(0.532)} = 9.28 \times 10^{-9}$

3. $\dfrac{(10^{-7})(3.142\ 31)}{(\pi)(68.2 \times 10^{-3})} = 1.47 \times 10^{-6}$

4. $\dfrac{(66.6)(0.937)}{(0.817 \times 10^4)(1.372)} = 5.95 \times 10^{-3}$

5. $\dfrac{(0.856 \times 10^{-3})(0.010\ 04)}{(83.2)(7\ 150\ 250)} = 1.44 \times 10^{-14}$

6. $\dfrac{(46.4)(93.2)}{(9.29)(0.501)} = 9.29 \times 10^2$

7. $\dfrac{(901\ 000)(0.156\ 2)}{(0.812)(0.836)(29.2 \times 10^{-5})} = 7.10 \times 10^8$

8. $\dfrac{(843\ 262)(987 \times 10^5)(\pi)}{(61.3 \times 10^8)(0.412)} = 1.04 \times 10^5$

9. $\dfrac{(176.3)(0.982)(0.811)}{(0.003\ 52)(5.62 \times 10^{14})} = 7.10 \times 10^{-11}$

10. $\dfrac{(608.6)(7.222)(32.1 \times 10^{17})}{(410\ 000)(6.390\ 5)(0.088\ 6)} = 6.08 \times 10^{16}$

Logarithms: $(\log = \log_{10};\ \ln = \log_e)$

1. $\log (83.2) = 1.92$

2. $\ln (0.853) = -1.59 \times 10^{-1}$

3. $\log (1.23 \times 10^5) = 5.09$

4. $\ln (9.15 \times 10^{-3}) = -4.69$

5. $\log (132.67) = 2.12$

6. $\log (14.2^{3.6}) = 4.15$

7. $\ln (86.3/52.7) = 4.93 \times 10^{-1}$

8. $\log (0.053) \log (3.92 \times 10^3) = -4.58$

9. Solve for x: $(1.053)^{6.31x} = 4.67 \rightarrow x = 4.73$

10. Solve for x: $x^{-2.3} = 0.567 \rightarrow x = 1.28$

Exponents $(e = 2.718\ 28 \ldots)$

1. $(0.806)^2 = 0.65$

2. $(53.2 \times 10^7)^2 = 2.83 \times 10^{17}$

3. $\sqrt{1.295} = 1.14$

4. $(0.005\ 67)^{\frac{1}{2}} = 7.53 \times 10^{-2}$

5. $(14.9)^2(23.568)^{\frac{1}{2}} = 1.08 \times 10^3$

6. $\sqrt[3]{460\ 052} = 7.72 \times 10^1$

7. $(37.2)^{4.32} = 6.09 \times 10^6$

8. $(8.37 \times 10^{-5})^{1/0.89} = 2.62 \times 10^{-5}$

9. $(3.81 \times 10^3)^{1.98}(0.823)^{-0.56} = 1.32 \times 10^7$

10. $(\pi)^2(6.805)^{-12.2} = 6.82 \times 10^{-10}$

11. $10^{6.35} = 2.24 \times 10^6$

12. $e^{-12.3} = 2.24 \times 10^{-6}$

13. $e^{65.2} = 2.07 \times 10^{28}$

14. $10^{-12.36} = 4.37 \times 10^{-13}$

15. $(10^{0.012})(10^{-15.3}) = 5.15 \times 10^{-6}$

16. $(e^{-0.562})(e^{3.3}) = 1.55 \times 10^1$

17. Solve for y:
$\ln y = 2.523; \qquad y = 1.246 \times 10^1$

18. Solve for y:
$\log (1/y) = 2.932; \qquad y = 1.169 \times 10^3$

Trigonometric Functions

1. $\sin 30° = 0.50$

2. $\sin(\pi/6) = 0.50$

3. $\cos 45° = 0.71$

4. $\tan(-137.2°) = 0.93$

5. $\csc 122.2° = 1.18$

6. $\cot(-36.35°) = -1.36$

7. $\sin 832° = 0.93$

8. $\sec 81.2° = 6.54$

9. $\tan(2.62\pi) = -2.53$

10. $\csc(-146°) = -1.79$

11. $\theta \text{ (degrees)} = \tan^{-1}(6.32) = 81.01°$

12. $\theta \text{ (radians)} = \text{arcsine}(-0.862) = 2.61$ radians

13. $\theta \text{ (degrees)} = \csc^{-1}(2.37) = 24.96°$

14. $(\sin 399°)(\cot 12.2°)(\cos 83°) = 0.355$

15. $\theta \text{ (degrees)} = \cos^{-1}(2/\sqrt{2^2 + 3^2}) = 56.31°$

Polar $\rightleftarrows$ Rectangular Conversion

1. $r = 6.21$
$\theta = 87.2°$ $\longrightarrow$ $x = 0.30$
$y = 6.20$

2. $r = 125.6$
$\theta = 283°$ $\longrightarrow$ $x = 28.25$
$y = -122.38$

3. $r = 0.020\ 4$
$\theta = 8.63 \text{ rad}$ $\longrightarrow$ $x = -1.43 \times 10^{-2}$
$y = 1.46 \times 10^{-2}$

4. $x = 17.2$
$y = -15.6$ $\longrightarrow$ $r = 23.2$
$\theta = -42.2°$

5. $x = -7.92 \times 10^{13}$
$y = -1.26 \times 10^{13}$ $\longrightarrow$ $r = 8.02 \times 10^{13}$
$\theta = -170.96°$

6. $x = 2.032$
$y = -16.32$ $\longrightarrow$ $r = 16.45$
$\theta = -1.47 \text{ rad}$

PROBLEMS

Work the following problems using your calculator and express the answers in scientific notation to two decimal places.

5.1 $\dfrac{(98.3)(6.23 \times 10^3)}{\pi} =$

5.2 $\dfrac{(0.005\ 67)(16.26)}{(3.005)(1.32 \times 10^{-5})} =$

5.3 $(59.17)(0.325)(0.005\ 52) =$

5.4 $(\pi)^2(63.2)^2(0.55)^{\frac{1}{2}} =$

5.5 $\sqrt[3]{0.082} =$

5.6 $(1.25 \times 10^3)^3(88.6)^3 =$

5.7 $(10^{-2.6})(e^{5.3}) =$

5.8 $(2.67 \times 10^{-5})(23.2^{0.811})(0.05^{3.2}) =$

5.9 $\sqrt[0.2]{16\ 302} =$

5.10 $(0.452)^{1/0.55} =$

5.11 $(\pi)(e^{3.2})^{2.1}(0.958)^{-2} =$

5.12 Solve for y:
   a. $3.12^y = 1.57$
   b. $(213\ 000)^{15.6y} = 5.89 \times 10^{65}$
   c. $\sqrt[0.7y]{28.92} = 123.26$
   d. $(1.67y)^{0.96} = 0.852$

5.13 $\log 0.563 =$

5.14 $\ln 9\ 902.6 =$

5.15 $\log(1.23 \times 10^{21}) =$

5.16 $(\ln 0.025)(\log 8.12) =$

5.17 Solve for y:
   a. $\ln(2.63y) = -1.22$
   b. $\log(1.11 \times 10^3) = 0.953y$
   c. $\ln(2.05/y) = 15.62$
   d. $\log(y^{2.81}) = 4.93$

5.18 $\sin 86.2° =$

5.19 $\tan(0.783\pi) =$

5.20 $\csc(-26°) =$

5.21 $\cos(-622°) =$

5.22 $\sec(0.873\pi) =$

5.23 $\dfrac{(\sin 172°)(\csc 172°)}{(\cos 42.1°)(\sec 42.1°)} =$

5.24 $\cos^{-1}(-0.21) =$ (°)

5.25 $\tan^{-1}(-121.2) =$ (rad)

5.26 $\tan(\sin^{-1} 0.52) =$

5.27 $\sec^{-1}(-1.26) =$ (rad)

5.28 $\sin^2 122° + \cos^2 105° =$

5.29 $\sqrt{15.2^2 + (20\cos 38°)^2} =$

5.30 $r = 701$ $x =$
      $\theta = 121°$ $y =$

5.31 $r = 2.53 \times 10^{-2}$ $x =$
      $\theta = \pi/5$ rad $y =$

5.32 $x = 19.54$ $r =$
      $y = -2.78$ $\theta =$ (°)

5.33 $x = -199.2$ $r =$
      $y = 88.2$ $\theta =$ (rad)

5.34 $\dfrac{(44.25)(0.08 \times 10^{12})(\sin 122°)}{(\sqrt{18.62})(96.7^{0.12})} =$

5.35 $\dfrac{(85.2^2 + 25.2^2)^{\frac{1}{2}}(e^{3.22})}{(67.15 \times 10^{-3})[1.23 + (0.21)(\tan 82.6°)]^{0.53}} =$

5.36 $\dfrac{(\pi^2)(\csc 22.2°)}{\sqrt[3]{99.9}} + (\cot 12°)(1.23 \times 10^{-2} + e^{-2.36})^{1.67} =$

5.37 $\dfrac{(23 \times 10^3)(0.197\ 2 \times 10^{-2})}{(76.1)^{1.4}(\sin 49.1°)} =$

5.38 $\dfrac{(48.2 - 6.3^2)(\sin 82°)}{(1.2 \times 10^7)(e^{22})\sqrt{100.5}} =$

5.39 $\dfrac{(\sqrt{\pi})[(0.07)^2(\sin 25°) + (30)(2.8)]^{1.25}}{(362)(\pi)(1.2 \times 10^{27})(5.2 \times 10^{-51})} =$

5.40 $\dfrac{(0.083)[\tan 136° + 4^{0.8}]^{3/2}}{(3.2 \times 10^{-3})(568.2)} =$

5.41 $\dfrac{(1.3^{2.5} + 5.37)(\sqrt{2.95} + \ln 5)}{(1.379 \times 10^5)(3.97 \times 10^3 - e^{10})} =$

5.42 $\dfrac{\sqrt[3]{193}}{\cot 145°} + \dfrac{1.5 \times 10^3}{0.062 \times 10^6} - \pi^3 =$

5.43 $\dfrac{(1.93 \times 10^{15})[(56.2)(e^{2.7}) - 3.2^5]^{0.05}}{(\pi)(\sin 48° - \cos 136°)} =$

5.44 $\dfrac{(1.37 \times 10^{-15})(10^{5.6})(\csc 135°)}{[(10)(\pi) + 4.6^{2.2}]^3} =$

5.45 $3.91 \times 10^2 + \dfrac{(126.7)(\tan 138°)}{\left(\cot \dfrac{3\pi}{4}\right)\sqrt{0.006}} =$

Dimensions and Units

CONTENTS

Historical Background
Dimensions
Unit Systems
International System of Units (SI)
Gravitational Unit Systems
Conversion of Units

Dimensions and units enable us to describe physical situations numerically, but, like languages, unit systems have proliferated, making technical communication throughout the world more complex. Students must understand several unit systems until the International System of Units (SI) finally becomes the worldwide standard, adopted not only in word but in actual practice.

HISTORICAL BACKGROUND

In the sixth chapter of the Book of Genesis, it is recorded that Noah built an ark to specifications:

> . . . The length of the ark shall be three hundred cubits, the breadth of it fifty cubits, and the height of it thirty cubits. (Genesis 6:15)

Historians record that the *cubit* was the ordinary unit of length among the ancient Hebrews. Originally, this length was the distance from the elbow to the fingertips. It is estimated that at the time of Noah, the cubit was about 0.45 meters (17–18 inches) in length;* later, at the time of Christ, the cubit was nearer 0.50 meters (20 inches) in length.

If we use a 0.45 meters per cubit conversion factor, then the ark was approximately 135 meters (150 yards) long, 22.5 meters (25 yards) wide, and 13.5 meters (45 feet) high. Thus, by standards common to us today, this ark was about one and one half football fields long, two thirds of a football field wide, and over four stories high. If these recorded ark dimensions are reasonably accurate, the building of this vessel must have been quite a feat at that time.

Other units of length were frequently used in this early period. The *meridian mile* can be traced back to 4000 B.C. It is estimated to have been equivalent to 4 000 cubits (about 1.85 kilometers) or 1 000 Egyptian *fathoms,* another early measure of length. Later, the Greeks and Romans set up their own standard mile as 1 000 double-step paces; a double step then was estimated to be about 1.5 meters long.

---

*Alternative units will often be placed in parentheses alongside the units currently under consideration. Exact conversions will usually be given, but sometimes, as in the examples given here, conversions will be only approximate.

Length was the first dimension for which standards evolved. Early units of length were related to human dimensions, such as the width of the finger, the handspan, the forearm, the foot, or the space of a single or double step. Since physical dimensions varied from one person to the next, as they do now, it is understandable that standards for measurements were so easily proliferated in our early history.

Quite often, the leaders of the period had to decide which standards would be applied by the builders, tradespeople, and merchants of that period. In A.D. 1324, King Edward II of England decreed that the official inch was the total length of three barleycorns taken from the middle of the ear and laid end to end. Later, in 1496, King Henry I established the yard as the distance from the tip of his nose to the end of the thumb on his outstretched arm. By the year 1500, the following units were well established as the units of length in England:

$$3 \text{ barleycorns} = 1 \text{ inch}$$
$$12 \text{ inches} = 1 \text{ foot}$$
$$3 \text{ feet} = 1 \text{ yard}$$
$$9 \text{ inches} = 1 \text{ span}$$
$$5 \text{ inches} = 1 \text{ ell}$$
$$5 \text{ feet} = 1 \text{ pace}$$
$$125 \text{ paces} = 1 \text{ furlong}$$
$$5\tfrac{1}{2} \text{ yards} = 1 \text{ rod}$$
$$40 \text{ rods} = 1 \text{ furlong}$$
$$8 \text{ furlongs} = 1 \text{ statute mile}$$
$$12 \text{ furlongs} = 1 \text{ league}$$

Another primary measurement of antiquity was **weight,** the force that a mass experiences as a result of the gravitational attraction of the earth. It is recorded in Genesis 23:16 that Abraham the prophet bought a burial field for "... 400 shekels [a weight measurement] of silver, current money with the merchants." We presume that the earliest standards for weight measurement were stones or other objects that merchants could heft to compare with the article being bargained for. Later, we know, crude scales were constructed so that purchases could be weighed against standardized weights. Scales are still the primary method for measuring weights today.

Closely akin to weight was **volume.** Interestingly enough, the Magna Carta in 1215 established the London quart as the specified unit of volume. Thus, we can trace the modern gallon (equivalent to 4 quarts) to this famous document.

As countries developed and began to exchange more and more commodities with one another, it became increasingly evident that "systems" of weights and measures were needed to make the exchange process less difficult. English-speaking countries adopted the fractional system of units we now recognize as the English system of units, still widely used throughout the United Kingdom (U.K.) and the United States. Later it was modified to allow decimal divisions also.

Meanwhile, a separate decimal system of units was evolving in France, eventually to become what we now know as the International System of Units (SI).* We chronologically trace a few of the more important events in this evolution as follows:

1791—France establishes the meter as the 10^{-7} part of the length of the line through Paris from the equator to the North Pole, that is, the meridional line through Paris.

1792—Over a seven-year period, Delambré obtains an estimate of the length of this line by surveying part of the meridional line between Barcelona and Dunkirk.

1799—A *standard* meter and kilogram are constructed and stored in France for reference.

1827—Babinet suggests that the wavelengths of light might be a more accessible standard of length.

1840—France outlaws all other units of measurement and makes the metric system the only legal system of units.

1866—Metric units are made legal in England and are accorded equal status with older, more established units.

1875—The seventeen-nation Metric Convention in Paris establishes the International Bureau of Weights and Measures and organizes the General Conference on Weights and Measures (CGPM)†

1889—The United Kingdom establishes the National Physical Laboratory and defines the meter as the distance between two marks on a platinum-iridium bar. At the same time, the kilogram is defined as the mass of a standard cylindrical bar of platinum-iridium.

1901—The United Kingdom establishes what is now known as the British Standards Institution, while the National Bureau of Standards is established in the United States.

1948—The CGPM moves to establish a practical system of units suitable for adoption by all signatories of the Metric Convention.

1959—The United States, United Kingdom, Canada, New Zealand, and South Africa define the yard to be 0.914 4 meter and the avoirdupois pound to be 0.453 592 237 kilogram.‡

1960—The name and abbreviation for the modified metric system of units (the International System of Units and SI, respectively) are adopted by the CGPM.

1971—The U.S. Department of Commerce issues its twelve-volume metric study interim report, which recommends conversion to the international metric system over a ten-year period.

---

*Simon Steven of Holland is thought to have originated the decimal system in the sixteenth century.
†CGPM stands for *Conférence Générale des Poids et Mesures*.
‡Avoirdupois is the measuring system used for all commodities except drugs, precious metals, and stones, which are figured in troy weight. We usually omit the term *avoirdupois* when we use weight units.

We have been slow to heed the recommendation made in 1971 by the U.S. Department of Commerce. Now, in the 1980s, industries and businesses still converse in more than one unit system. Clearly, this "bilingual" requirement places a significant burden on all of us. In addition, it constrains our ability to engage in foreign trade (importation and exportation). Nevertheless, time is on the side of SI. Even though resistance to SI in the United States continues, it seems inevitable that we will convert to this more consistent set of units.

The necessary emphasis is already being placed on SI in our primary and secondary schools and in our colleges and universities. Many engineering courses now use metric units as the primary units and English units or American engineering units as secondary units. The emphasis in this text is consistent with this trend. Metric units are the medium in most numerical calculations and are required in most of the problems. However, enough emphasis is placed on nonmetric units so that students can become reasonably bilingual. Often, alternative units are placed in parentheses alongside the primary units under consideration to facilitate this process.

DIMENSIONS

Dimensions are unique quantities that describe in a physical way some characteristic of a system or entity. In simpler terms, a dimension is what we use to evaluate physical phenomena. The most fundamental of these dimensions can be combined through natural laws or mathematical relationships to describe most physical occurrences. Thus, dimensions can be attributed to fundamental qualities like force, mass, length, and time; these are called **fundamental** or **primary dimensions.** An abbreviated way of identifying fundamental dimensions is by using a single letter like L for length, M for mass, t for time, and T for temperature. Table 6.1 lists those quantities that are usually considered fundamental, depending on which unit system is referenced. As we will see, the choice of fundamental dimensions varies among unit systems.

Derived dimensions are some combination of the fundamental dimensions. For example, the volume of the rectangular parallelepiped in Figure 6.1

TABLE 6.1 Fundamental Dimensions

Fundamental dimension	Symbol
Mass	M
Length	L
Time	t
Force	F
Temperature	T
Charge	Q
Electric current	I
Molecular substance	n
Luminous intensity	I_ℓ

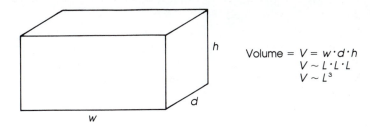

Volume $= V = w \cdot d \cdot h$
$V \sim L \cdot L \cdot L$
$V \sim L^3$

FIGURE 6.1

is simply a product of the width (L), depth (L), and height (L). It has a derived dimension that is a product of three length dimensions, that is, L^3. Similarly, *speed* is a change in distance (L) divided by a change in time (t), giving the derived dimension L/t or $L \cdot t^{-1}$.

In contrast, there are quantities that have no fundamental dimensions; these are called **dimensionless quantities.** Some of the more obvious examples are ratios of like quantities and coefficients, like the coefficient of friction. A **conversion factor,** which is used to convert from one set of units to another, is an example of a dimensionless ratio. Some common conversion factors are 12 inches per foot, 60 seconds per minute, 100 centimeters per meter, and 1 000 grams per kilogram. We should point out that an angle is considered a dimensionless quantity, although there is some controversy over this point.

Dimensional Analysis

Engineers and technologists manipulate equations of various kinds. These equations may be derived from experimental data or may be mathematical counterparts of fundamental laws. They may be composed of transcendental functions (like trigonometric functions), they may be simple algebraic equations, or they may involve complex differential and integral relationships. Regardless of the simplicity or complexity of the manipulations involved, it is necessary that the equations be **dimensionally consistent** or **homogeneous** at every stage in the manipulation. If they are not, either the manipulations are in error or the equation was dimensionally inconsistent to begin with.

Example 6.1

The equation that follows describes the water pressure p at a depth h below the surface of a body of water. What are the fundamental dimensions of the constant k, and what does this constant physically represent if p_o is the atmospheric pressure above the water?

$$p = p_o + kh$$

SOLUTION
The fundamental dimensions of pressure are F/L^2, and the fundamental dimension of depth h is L. Comparing the fundamental dimensions of each

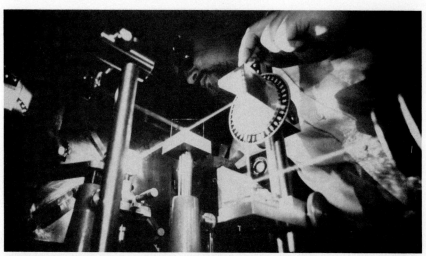

Length, time, and angular orientation are dimensional concepts important to most engineering experiments, like the splitting and reflecting of the light beam shown. (Courtesy of Honeywell)

term on each side of the pressure equation, the equation for p reduces to the dimensional equation

$$\frac{F}{L^2} \sim \frac{F}{L^2} + k \cdot L$$

Dimensional homogeneity requires that the fundamental dimensions on both sides of the equation be identical. Thus, the fundamental dimensions of kh must reduce to that of pressure. This requires that $kL \sim F/L^2$. Dividing both sides by L,

$$k \sim \frac{F}{L \cdot L^2} = \frac{F}{L^3}$$

Thus, the fundamental dimensions of k are F/L^3. We conclude that k represents the specific weight of water, often denoted by the symbol γ, since k has dimensions of force (the force being the weight of water) per volume of substance.

Dimensional consistency is only a *necessary* condition for equation validity; it is not a *sufficient* condition. In other words, how accurately an equation models some physical phenomenon depends primarily on the assumptions made in the modeling process, such as parameter selection, discarding of second-order effects, and so forth. In cases where equations are generated from data that are derived by experiment, it is obvious that the experiment must be conducted properly if the data are to have any representative meaning. In conclusion, then, equations must be dimensionally consistent; nevertheless, dimensional consistency cannot guarantee that the equations are representative of any physical phenomena.

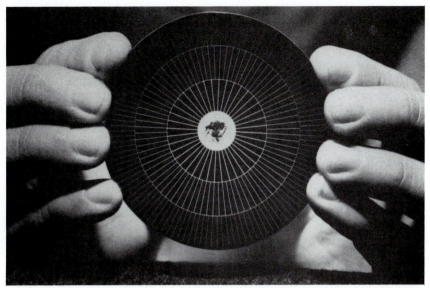

Although not a fundamental dimension, the angle is an important physical concept; witness the design symmetry of the photovoltaic cell shown. (Courtesy of Lockheed)

Example 6.2

An engineer performs an experiment to measure the dynamic viscosity μ of a liquid flowing steadily between two points in a pipe. **Viscosity** is a measure of the frictional resistance of a fluid to flow. The engineer solves the following equation for μ:

$$\frac{p_1}{\gamma} + Z_1 = \frac{p_2}{\gamma} + Z_2 + \frac{32\mu V}{\gamma D^2}$$

where

p_1, p_2 = pressures at two different pipe locations
Z_1, Z_2 = heights of the pipe above ground at these two locations
γ = specific weight of the fluid
V = average velocity
D = pipe inside diameter
μ = dynamic viscosity

Given that viscosity has fundamental dimensions of $F \cdot t / L^2$, is the engineer applying a dimensionally consistent equation?

SOLUTION
Examine the fundamental dimensions of each segregated term in the equation. *For dimensional consistency, each segregated term must have the same fundamental dimensions when reduced to its simplest form.* Given that pres-

TABLE 6.2

Term(s)	Dimensional reduction	Fundamental dimension(s)
$p_1/\gamma,\ p_2/\gamma$	$\dfrac{F/L^2}{F/L^3} = \dfrac{F \cdot L^3}{F \cdot L^2} \qquad =$	L
$Z_1,\ Z_2$	$L \qquad\qquad =$	L
$\dfrac{32\mu V}{\gamma D^2}$	$\dfrac{(F \cdot t/L^2)(L/t)}{(F/L^3)(L)^2} = \dfrac{F \cdot t \cdot L^4}{F \cdot t \cdot L^4} =$	Dimensionless!

sure is a force per area, specific weight is a weight per volume, and V is a velocity, the fundamental dimensions of each term in this equation can be reduced to their simplest form as shown in Table 6.2.

CONCLUSION
Since all terms of the equation do not have the same fundamental dimensions, the equation is not dimensionally consistent. It is incorrect in its present form.

DON'T STOP NOW
It would be a mistake to leave this problem now. Suppose you are the engineer who performed this fluids experiment only to discover that the value you determined for μ was not reasonable—at least on the basis of your previous experience with values for μ. As an engineer, you would check your calculations before assuming that the experiment was not conducted properly. After obtaining the same incorrect answer, you would then check the form of the equation by applying dimensional analysis. Upon discovering the dimensional inconsistency in the term $32\mu V/\gamma D^2$, you would probably realize that a length is missing from the last term. In all likelihood, you would recognize that this missing length is the length of the pipe between the two points at which the pressure is measured. Letting L_p be this length, the form of the final term would then be corrected to $32\mu L_p V/\gamma D^2$.

Reduction of Fundamental Dimensions
Fundamental dimensions can be canceled by dividing but not by subtracting, because fundamental dimensions have no numerical significance attached to them, only physical significance. This can be illustrated by the following example.

Example 6.3

Determine the fundamental dimensions for q given that the following heat transfer equation is dimensionally consistent.

$$q = \frac{kA(T_2 - T_1)}{l}$$

where

k = coefficient of thermal conductivity in units of watts per meter per kelvin

A = cross-sectional area in square meters

l = a length in meters

T_1, T_2 = absolute temperatures in kelvin.

SOLUTION

A watt is a unit of power, and power has the fundamental dimensions of $F \cdot L/t$. Thus, k has the fundamental dimensions

$$k \sim \frac{F \cdot L/t}{L \cdot T} \sim \frac{F}{t \cdot T}$$

The fundamental dimensions for q can now be determined as:

$$q \sim \frac{(F \cdot t^{-1} \cdot T^{-1})(L^2)(T - T)}{L} \qquad \text{Note that } (T - T) \sim T$$

$$\sim \frac{F \cdot L^2 \cdot T}{t \cdot T \cdot L}$$

$$\sim F \cdot L/t$$

Thus, q has the fundamental dimensions of power, in other words, a force times a distance per time. In this case, q physically represents the rate of heat energy being conducted through a material of thickness l and cross-sectional area A. Note that in the dimensional reduction, $(T - T)$ has a resultant fundamental dimension of T, not zero dimensions. It makes no sense to subtract fundamental dimensions and end up with no dimension at all.

Derived Dimensions

As mentioned earlier, derived dimensions are some combination of the fundamental dimensions. There are a multitude of them in engineering. In fact, engineers will work with derived dimensions as often as with fundamental dimensions. Some common derived dimensions are presented in Table 6.3 and defined in the simplest of terms where necessary.

Although not listed in Table 6.3, some fundamental dimensions may be considered derived dimensions, depending on which unit system is being used. For example, in SI, charge is a derived dimension, whereas current is the fundamental dimension. In the English system of units, charge is fundamental and current is derived. Similarly, in SI, mass is fundamental, whereas force is derived; this contrasts with the English system of units, in which force is considered a fundamental dimension and mass a derived dimension.

Another problem of having different fundamental dimensions in different unit systems is that it can confuse the dimensional analysis somewhat. For example, the potential difference in the English system, as shown in Table 6.4, has derived dimensions of $F \cdot L/Q$. Yet, if we had considered current and

TABLE 6.3 Some Derived Dimensions

Quantity	Definition	Dimensions
Acceleration	Rate of change of velocity	L/t^2
Angular acceleration	Rate of change of angular velocity	t^{-2}
Angular velocity	Rate of change of angular position	t^{-1}
Area	Space contained within a surface element	L^2
Capacitance (electric)	Charge per potential difference	$Q^2/(F \cdot L)$
Concentration	Number of moles per volume	n/L^3
Density (mass)	Mass per volume	M/L^3
Energy	Capacity for doing work	$F \cdot L$
Frequency	Number of cycles per time	t^{-1}
Inductance (electric)	Voltage per rate of change of current	$F \cdot L/I^2$
Potential difference (electric)	Work per charge (also called voltage)	$F \cdot L/Q$
Power	Rate of work or change in energy	$F \cdot L/t$
Pressure	Force per area	F/L^2
Resistance (electric)	Voltage drop divided by the current	$F \cdot L/(t \cdot I^2)$
Specific heat	Heat energy per unit mass per degree	$F \cdot L/(M \cdot T)$
Specific volume	Volume per mass	L^3/M
Specific weight	Weight per volume	F/L^3
Speed (scalar)	Rate of change of position	L/t
Stress	Force per area	F/L^2
Torque	Force times a perpendicular distance	$F \cdot L$
Velocity (vector)	Rate of change of position	L/t
Volume	Space within a three-dimensional element	L^3
Work	Force times a distance	$F \cdot L$

TABLE 6.4 Derived Dimensions in Different Unit Systems

Quantity	Derived dimensions	
	SI	*English unit system*
Capacitance	$I^2 \cdot t^4/(M \cdot L^2)$	$Q^2/(F \cdot L)$
Charge	$I \cdot t$	$Q*$
Current	$I*$	Q/t
Force	$M \cdot L/t^2$	$F*$
Mass	$M*$	$F \cdot t^2/L$
Potential difference (voltage)	$M \cdot L^2/(I \cdot t^3)$	$F \cdot L/Q$
Work or energy	$M \cdot L^2/t^2$	$F \cdot L$

*Denotes a fundamental dimension in that unit system.

mass as fundamental, the dimensions of the potential difference could have been represented as $M \cdot L^2/(I \cdot t^3)$. Nevertheless, it is not difficult to make the conversion to this dimensional form from the other. You first relate F to M, L, and t through Newton's second law and then Q to I and t by the definition of current as the rate of change of charge to obtain the form listed in the table.

Three fundamental laws relating dimensions are so important to the derivation of derived dimensions that we present them now.

NEWTON'S SECOND LAW   If a particle is subjected to unbalanced forces, the particle will have an acceleration proportional to the magnitude of the resultant of the unbalanced forces and in the direction of the resultant force.

$$F = ma$$

where
   F = resultant force on particle
   a = acceleration of particle
   m = mass, the constant of proportionality

COULOMB'S LAW   The force between two fixed charged particles is directly proportional to the product of their charges and inversely proportional to the square of the distance between them. In symbols,

$$F = k_e \frac{q_1 q_2}{r^2}$$

where
   F = force of attraction or repulsion in newtons
   q_1 = charge on particle 1 in coulombs
   q_2 = charge on particle 2 in coulombs
   r = distance separating particles in meters
   k_e = constant of proportionality, 9×10^9 newton meters squared per coulomb squared

AMPERE'S LAW   The electric currents in parallel wires of infinite length and negligible cross section produce a force between these two conductors *per length* of wire that is directly proportional to the product of the currents and inversely proportional to the minimum distance between the conductors. In symbols,

$$F_\ell = k_m \frac{I_1 I_2}{r}$$

where
   F_ℓ = force of attraction or repulsion per length of conductor in newtons per meter
   I_1 = current in conductor 1 in amperes
   I_2 = current in conductor 2 in amperes
   r = distance separating conductors in meters
   k_m = constant of proportionality, 2×10^{-7} newtons per ampere squared

TABLE 6.5 Some Units for Fundamental Dimensions

Fundamental dimension	Representative units
Force (F)	Newton, pound force, dyne
Length (L)	Meter, centimeter, inch, foot, mile
Mass (M)	Gram, kilogram, slug, pound mass
Time (t)	Second, minute, hour, year, decade

UNIT SYSTEMS

Whereas fundamental dimensions are used to describe the "why," it takes **units** to answer the question "how much" because units "quantify" the fundamental dimensions. Examples of some possible units for several fundamental dimensions are listed in Table 6.5.

Because of the proliferation of units, attempts were made to organize the various units into manageable unit "systems"; these attempts only resulted in the proliferation of additional "systems of units." In fact, twelve unit systems have been proposed in the past, ten of which have been frequently used. The four systems most often used in the United States are

1. The International System of Units (SI).
2. The centimeter-gram-second (CGS) system of units.
3. The English system of units.
4. The American engineering or just engineering system of units.

These systems are most easily distinguished by whether they are absolute unit systems or gravitational unit systems.

Absolute unit systems are independent of gravitational effects for the definition of their units; in other words, the concept of weight does not influence the definition of the units. Perhaps it is helpful to recall that an object's weight is the product of its mass and the acceleration of gravity, which can be determined using Newton's law of gravitation. Thus, we can conclude that in absolute unit systems, *mass* is the fundamental dimension rather than force. Both the SI and CGS systems of units fit into this category.

Gravitational unit systems consider force more fundamental than mass and use some standard weight to define the units of force. Both the English and the engineering systems of units are gravitational unit systems by virtue of the definition of the pound force, which we will give shortly.

INTERNATIONAL SYSTEM OF UNITS (SI)

The International System of Units is a decimal unit system that has seven fundamental or **base** dimensions, as they are called in SI (Table 6.6). Corre-

TABLE 6.6 SI Base Units

Fundamental quantity	Dimensional symbol	Unit	Unit symbol
Electric current	I	ampere	A
Length	L	meter	m
Luminous intensity	I_ℓ	candela	cd
Mass	M	kilogram	kg
Molecular substance	n	mole	mol
Temperature	T	kelvin	K
Time	t	second	s

TABLE 6.7 SI Supplementary Units

Quantity	Symbol	Unit	Unit symbol
Plane angle	Greek symbols (α, β, γ, etc.)	radian	rad
Solid angle	Greek symbols (α, β, γ, etc.)	steradian	sr

spondingly, the units for these base dimensions are called base units. In addition, there are two **supplementary** units for angles (Table 6.7).

Units are fundamentally different from dimensions in that they are quantitative, whereas dimensions are qualitative. Consequently, units must be specified according to some measurement **standard.** For example, in the past, the standard for length was the distance between two marks on a metal bar kept at some fixed location and at a reasonably constant temperature. Unfortunately, this kind of standard is not readily accessible as a reference. Newer standards for units (except for mass) employ modern technology and fundamental physical phenomena for their definitions—phenomena that can be reconstructed at any convenient location given the prerequisite technology. The current SI unit definitions demonstrate this reliance on physical phenomena and modern technology.*

LENGTH The base unit of length is the **meter** (m), which is defined as 1 650 763.73 wavelengths in a vacuum of the radiation corresponding to the unperturbed transition between the energy levels $2P_{10}$ and $5d_5$ of the krypton-86 atom (Figure 6.2). The wavelength of this orange-red line is $6\,057.802 \times 10^{-10}$ meter.

MASS The base unit of mass is the **kilogram** (kg), currently the mass of a platinum-iridium alloy cylinder kept by the International Bureau of Weights and Measures at Paris, France (a duplicate copy is maintained in Washington, D.C.).

---

*The following definitions are from the National Bureau of Standards Special Publication 330.

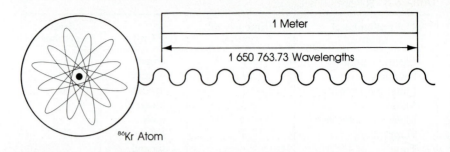

FIGURE 6.2

TIME   The base unit of time is the **second** (s). It is the time duration of 9 192 631 770 periods of the radiation of the transition between the "hyperfine" levels (energy levels that are only slightly separated) of the ground state of the cesium-133 atom.

Figure 6.3 is a schematic diagram of a beam spectrometer or "clock." Only those atoms whose magnetic moments are "flipped" in the transition region reach the detector. When 9 192 631 770 oscillations have occurred, the clock indicates that 1 second has passed.

ELECTRIC CURRENT   The base unit of electric current is the **ampere** (A). It is defined from Ampère's law as the constant current that, if maintained in two straight parallel conductors of infinite length and negligible cross-sectional area that are placed exactly 1 meter apart in a vacuum, will produce between them a force of 2×10^{-7} newton per meter.

TEMPERATURE   The base unit of temperature is the **kelvin** (K). Note that it *should not* be referred to as "degree Kelvin." It is defined as the fraction

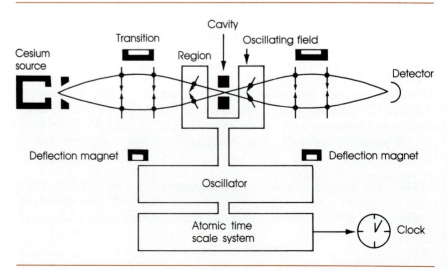

FIGURE 6.3

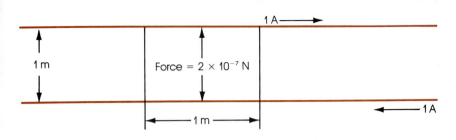

FIGURE 6.4

1/273.16 of the thermodynamic temperature at the triple point of water, that is, that point where water exists in three phases simultaneously: solid, liquid, and gas (here the gas is a water vapor). A **thermodynamic** temperature is an absolute temperature scale independent of the properties of any substance. This is in contrast to a temperature measured in degrees Celsius, a relative scale based on the freezing and boiling points of water at standard atmospheric pressure. The four most common temperature scales are shown in Table 6.8.

Note that the relationships between these temperature scales are easily established, since there is 100° between freezing and boiling on both the absolute and relative international scales and 180° between the same two points on the English scales. Remembering to add in the 32°F temperature shift between the Celsius and Fahrenheit scales, we can arrive at the following relationships:

$$T(K) = T(°C) + 273.16 = \frac{5}{9}T(°R)$$

$$T(°R) = T(°F) + 459.69 = \frac{9}{5}T(K)$$

$$T(°F) = \frac{9}{5}T(°C) + 32$$

$$T(°C) = \frac{5}{9}[T(°F) - 32]$$

TABLE 6.8 Absolute and Relative Temperature Scales

Condition	International temperature scales		English temperature scales	
	Kelvin (K)	Celsius (°C)	Rankine (°R)	Fahrenheit (°F)
Absolute zero	0.0	−273.15	0.0	−459.69
Water freezing point	273.15	0.00	491.69	32.00
Water triple point	273.16	0.01	491.71	32.02
Water boiling point	373.15	100.00	671.69	212.00
	SI absolute	Relative	Absolute	Relative

MOLECULAR SUBSTANCE The base unit of molecular substance is the **mole** (mol). It is defined as the amount of substance of a system that contains as many elementary entities as there are atoms in 0.012 kilogram of carbon-12. The mole is often related to Avogadro's constant, 6.023×10^{23}, which is the number of molecules in any mole of a substance. A more basic definition of the mole is that it is the gram molecular weight of that material. For example, carbon has a molecular weight of 12.011 2. A mole of carbon, by definition, would then weigh approximately 12 grams or 0.012 kilogram.

LUMINOUS INTENSITY The base unit of luminous intensity is the **candela** (cd). It is defined as the intensity, in the perpendicular direction, of 1/600 000 of a square meter of a blackbody at the temperature of freezing platinum (2 045 K) under a pressure of 101 325 newtons per square meter.

PLANE ANGLE The supplementary unit of the plane angle is the **radian** (rad). It is the angle formed between two radii of a circle and subtended by an arc whose length is equal to the radius.

SOLID ANGLE The supplementary unit of the solid angle is the **steradian** (sr). The steradian is the solid angle with its vertex at the center of a sphere that is subtended by an area of the spherical surface equal to that of a square with sides equal in length to the radius (Figure 6.5).

SI Derived Dimensions

Many of the units for SI derived dimensions have special names, even though they can be related back to the base units through definitions or fundamental laws. We list some pertinent units in Table 6.9 along with their names and

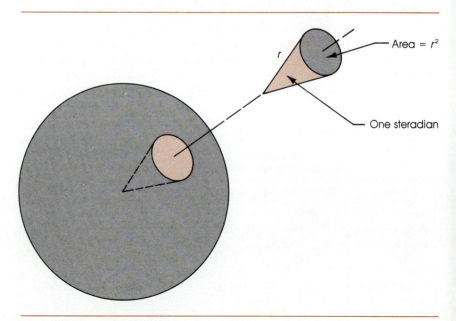

FIGURE 6.5

TABLE 6.9 SI Derived Dimensions and Units

Quantity	Unit	New name	Symbol
Acceleration	m/s^2	*	*
Angular acceleration	rad/s^2	*	*
Angular velocity	rad/s	*	*
Area	m^2	*	*
Capacitance (electric)	A·s/V	farad	F
Charge (electric)	A·s	coulomb	C
Concentration	n/m^3	*	*
Density (mass)	kg/m^3	*	*
Electric potential difference	W/A or J/C	volt	V
Energy (work)	N·m	joule	J
Force	kg·m/s^2	newton	N
Frequency	cycle/s	hertz	Hz
Inductance (electric)	V·s/A	henry	H
Power	J/s	watt	W
Pressure (stress)	N/m^2	pascal	Pa
Resistance (electric)	V/A	ohm	Ω
Specific heat	J/(kg·K)	*	*
Specific volume	m^3/kg	*	*
Specific weight	N/m^3	*	*
Speed (velocity)	m/s	*	*
Torque	N·m	*	*
Volume	m^3	*	*

*No new name or abbreviation is specified.

symbols. In addition, we consider several important units separately, to demonstrate how these units were derived or defined.

FORCE   The unit of **force** is the **newton** (N). It is derived from Newton's second law as the force necessary to accelerate a 1-kilogram mass 1 meter per second squared (1 m/s^2). This derivation follows by applying a to-be-defined force to a 1-kilogram mass and accelerating it 1 meter per second squared.

$$F = ? \quad \xrightarrow{} \boxed{m = 1 \text{ kg}} \quad \xrightarrow{} \quad a = 1 \text{ m/s}^2$$

From Newton's second law,

DEFINITION $F = 1 \text{ kg·m/s}^2 = 1 \text{ newton} = 1 \text{ N}$

WORK/ENERGY   The unit of **work** or **energy** is the **joule** (J). It is derived from the work of a constant 1-newton force moving through a distance of 1 meter, as shown in Figure 6.6. By definition,

$$\text{Work} = \text{Force·Distance} = 1 \text{ N·1 m}$$

DEFINITION   Work $= 1 \text{ N·m} = 1 \text{ joule} = 1 \text{ J}$

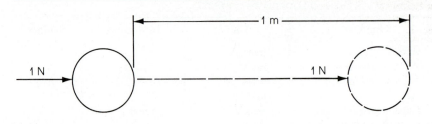

FIGURE 6.6

POWER   The unit of power is the **watt** (W). **Power** is the rate at which work is being done or energy is being transferred. If 1 joule of energy is transferred in 1 second, then the average power is 1 watt.

$$\text{Power} = \text{Energy/Time} = (1 \text{ J})/(1 \text{ s})$$

DEFINITION $\text{Power} = 1 \text{ J/s} = 1 \text{ watt} = 1 \text{ W}$

PRESSURE   The unit of pressure is the **pascal** (Pa). **Pressure** is a force per area given that the force is distributed over the area and normal to it. A pascal is the force of 1 newton over an area of 1 meter squared.

$$\text{Pressure} = \text{Force/Area} = (1 \text{ N})/(1 \text{ m}^2)$$

DEFINITION $\text{Pressure} = 1 \text{ N/m}^2 = 1 \text{ pascal} = 1 \text{ Pa}$

CHARGE (ELECTRIC)   The unit of charge is the **coulomb** (C). It is the quantity of electricity moved during 1 second by a current of 1 ampere. The coulomb is also defined as the charge of $6.241\ 96 \times 10^{18}$ electrons.

DEFINITION $1 \text{ A} \cdot \text{s} = 1 \text{ coulomb} = 1 \text{ C}$

POTENTIAL DIFFERENCE (ELECTRIC)   The unit of potential difference (or voltage or electromotive force) is the **volt** (V). A potential difference of 1 volt will cause a current flow of 1 ampere between two points in a circuit when the power dissipated between those two points is 1 watt. Note that the volt is also equal to 1 joule per coulomb.

DEFINITION $1 \text{ W/A} = 1 \text{ volt} = 1 \text{ V}$

Special SI Rules
Recommended practices for using SI units follow. They should be adhered to whenever possible.

1.   Decimal numbers smaller than 1 should be preceded with a zero.

YES   0.086
NO   .086

2. Thousands and thousandths should be separated by a space.

YES	1 693 585	YES	0.083 675
NO	1,693,585	NO	0.083675

In numbers of only four digits it is acceptable to omit the spaces.

3585 *or* 3 585   OKAY

0.0301 *or* 0.030 1   OKAY

3. The preferred powers of 10 are the third powers.

YES 10×10^3

NO 1×10^4

4. The acceptable SI word prefixes to be used in place of the powers of 10 are listed in Table 6.10. It is acceptable but not preferable to use the prefixes *hecto* (10^2), *deka* (10^1), *deci* (10^{-1}), and *centi* (10^{-2}).

TABLE 6.10 SI Unit Prefixes

Power of 10	Prefix name	Symbol
10^{15}	peta	P
10^{12}	tera	T
10^9	giga	G
10^6	mega	M
10^3	kilo	k
10^2	hecto	h
10^1	deka	da
10^{-1}	deci	d
10^{-2}	centi	c
10^{-3}	milli	m
10^{-6}	micro	μ
10^{-9}	nano	n
10^{-12}	pico	p
10^{-15}	femto	f

Symbols and Abbreviations

1. Unit symbols that are *abbreviated* from a *proper name* have the *first letter capitalized;* otherwise, they are written in lowercase letters. If the names of units are not abbreviated, then they are written in lowercase letters (unless at the beginning of a sentence). For example:

YES   pascal, P; meganewtons, MN

NO   Pascal, p; Meganewtons, mn

2. Abbreviated symbols are never made plural by the addition of an *s*. For names of units not abbreviated, it is permissible to form plurals according to the usual rules of sentence construction.
3. Units are spelled out if there is any possibility of confusion.

Mathematical Operations

1. The preferred method of indicating multiplication among SI unit symbols is by using the raised dot. Note that a space is always left between the number and the symbol.

$$7 \text{ kg} \times 6 \text{ m} = 42 \text{ kg} \cdot \text{m}$$

When writing out the product of unit names, it is preferred that the names be separated by a space or hyphen and that the powers also be written out rather than numerically displayed.

YES	newton meter	YES	kilogram meter cubed
NO	newton · meter	NO	kilogram meter³

2. Quotients of unit symbols can be indicated by any of the following forms:

$$\frac{\text{kg}}{\text{m}^3} \quad or \quad \text{kg/m}^3 \quad or \quad \text{kg} \cdot \text{m}^{-3} \quad \text{OKAY}$$

Quotients of unit names should be written using "per" rather than "/".

YES	kilogram per meter
NO	kilogram/meter

SI Exceptions

Although, strictly speaking, they do not belong to SI, there are several units that can be used in conjunction with SI units. Table 6.11 lists some of these units along with their conversion factors.

TABLE 6.11 Units Acceptable for Use with SI

Quantity	Name	Symbol	Conversion factors
Plane angle	degree	°	1° = 57.3 rad
Temperature	degree Celsius	°C	T(°C) = T(K) − 273.16
Time	minute	min	1 min = 60 s
	hour	h	1 h = 60 min
	day	d	1 d = 24 h
	week	week	1 week = 7 d
Volume	liter	ℓ or L	1 ℓ = 1 000 cm³

TABLE 6.12 CGS Units

Quantity	Unit	New name	Abbreviation
Acceleration	cm/s^2	*	*
Force	$g \cdot cm/s^2$	dyne	dyne
Length	centimeter	*	cm
Mass	gram	*	g
Power	erg/s	*	*
Pressure†	$dyne/cm^2$	*	*
Time	second	*	s
Velocity	cm/s	*	*
Work or energy	$dyne \cdot cm$	erg	erg

*No new name and/or abbreviation is specified.
†A metric unit for pressure that is still in common use is the bar. 1 bar is 10^6 dyne/cm^2.

CGS Unit System

Another absolute unit system that is still being used is the centimeter-gram-second unit system, or, as usually abbreviated, CGS (Table 6.12). Many of the calculations and physical constants in chemistry and electrical engineering are in CGS units. Since CGS units are metric units, they are easily related to SI units. The electrical units are defined similarly to those in SI.

GRAVITATIONAL UNIT SYSTEMS

The English and engineering gravitational systems of units are quite similar, the only real difference being in the way that mass is defined (Table 6.13). Remember that in gravitational unit systems, the force is considered the fundamental quantity and mass the derived quantity. In both unit systems the pound force, abbreviated lb_f, is the force unit. We define the pound force as *the force required to give a mass of 1 pound mass (lb_m) an acceleration of 32.173 2 feet per second squared,* in other words, the acceleration of gravity

TABLE 6.13 English and Engineering Units

Quantity	Unit	Symbol
Acceleration	feet per second squared	ft/s^2
Force	pound force	lb_f
Length	foot	ft
Mass (engineering)	pound mass	lb_m
Mass (English)	slug	slug
Power	foot-pounds per second	$ft \cdot lb_f/s$
Pressure	pounds per square foot	lb_f/ft^2
Time	second	s
Velocity	feet per second	ft/s
Work or energy	foot-pound	$ft \cdot lb_f$

at some standard position on the surface of the earth.* In the United States, the pound mass is defined as 453.592 427 7 grams; thus, it is related back to the metric standard for mass.

We define the units of mass in the English system of units by using Newton's second law and by applying a 1-pound force to a particular body having a mass that is accelerated by 1 foot per second squared. This mass is defined to be 1 **slug** in the following manner:

$$F = 1 \text{ lb}_f \quad \boxed{m = ?} \quad a = 1 \text{ ft/s}^2$$

Solving for m from $F = ma$, we have

$$m = F/a$$

and substituting for F and a, we get

$$m = (1 \text{ lb}_f)/(1 \text{ ft/s}^2)$$

DEFINITION $\qquad\qquad m = 1 \text{ lb}_f \cdot s^2/\text{ft} = 1 \text{ slug}$

In the engineering system of units, the mass is the pound mass and the force is the pound force. This poses a problem, since we have previously related the pound force to a mass in slugs in the English system of units. This necessitates a conversion factor that converts the mass in pound mass to that in slugs. To determine the value of this conversion factor, we apply the fundamental definition of a pound force and Newton's second law, rewritten to accommodate this conversion factor. Since a conversion factor is a non-dimensional quantity of unit magnitude, it does not affect the validity of this law.

$$F = 1 \text{ lb}_f \quad \boxed{m = 1 \text{ lb}_m} \quad a = 32.173\ 2 \text{ ft/s}^2$$

To determine this conversion factor, which we will identify by k, we rewrite $F = ma$ as $F = kma$ so that

$$F = 1 \text{ lb}_f = k(1 \text{ lb}_m)(32.173\ 2 \text{ ft/s}^2)$$

Solving for the conversion factor k and approximating 32.173 2 by 32.2,

$$k = (1 \text{ lb}_f \cdot s^2)/(32.2 \text{ lb}_m \cdot \text{ft})$$

Using the previous definition for a slug, that is, 1 slug = 1 $\text{lb}_f \cdot s^2/\text{ft}$, we recognize that k is simply the conversion factor from pounds mass to slugs.[†]

---

*In SI, the acceleration of gravity is approximately 9.807 m/s^2.

[†] Sometimes this conversion factor is expressed by the reciprocal constant g_c, where g_c = 32.2 $\text{lb}_m \cdot \text{ft}/(\text{lb}_f \cdot s^2)$.

$$k = \frac{1 \text{ lb}_f \cdot s^2/ft}{32.2 \text{ lb}_m} = 1 \text{ slug}/32.2 \text{ lb}_m$$

or

$$1 \text{ slug} = 32.2 \text{ lb}_m$$

If we relate the definition of the pound force to an acceleration equivalent to that of gravity, as was done in the definition of pound mass, then the weight of an object on or near the surface of the earth expressed in pounds force will always have a *numerical* value equal to the *numerical* value of the mass in pounds mass. This has caused considerable confusion. People who do not understand the difference between mass and weight often treat them identically. And to complicate matters further, the force unit will sometimes be abbreviated as lb rather than lb_f.

Another common mistake is to say something weighs so many kilograms; this is incorrect, since the kilogram is a unit of mass like the pound mass. We illustrate these differences by the following example.

Example 6.4

One of the purposes of the Apollo program was to bring back rock samples from the moon's surface. If the rock samples weighed 30 lb_f on the moon, where the acceleration of gravity is one sixth that of earth, what is the total mass of the samples in pounds mass? What would their weight be on earth in pounds force and in newtons?

SOLUTION

Apply Newton's second law, where $F = 30$ lb_f, m is in pounds mass, and g_m is the acceleration of gravity on the moon, which is approximately one sixth that of gravity on the earth. Remember to include the appropriate conversion factor k in the equation.

$$F = kmg_m$$

In this case F is the weight of rock samples on the moon and g_m is the moon's gravitational acceleration. Substituting,

$$F = 30 \text{ lb}_f = \left(\frac{1 \text{ lb}_f \cdot s^2}{32.2 \text{ lb}_m \cdot ft}\right)(m)\left(\frac{32.2 \text{ ft}/s^2}{6}\right)$$

Canceling units and solving for m, we get

$$m = 180 \text{ lb}_m$$

On the earth the weight of these rocks would be 180 lb_f. Using the 4.448 N/lb_f conversion factor, the weight of these rocks on the earth, in newtons, can be calculated as

$$W = (180 \text{ lb}_f)(4.448 \text{ N/lb}_f)$$
$$= \underline{800.6 \text{ N}}$$

CONVERSION OF UNITS

Conversion of units is not very difficult, since all that is necessary is a conversion **string** between the current units and the desired units. To illustrate, we will convert 6 inches (in.) to kilometers; but first we must recall the following "fundamental" conversions:

$$1 \text{ in.} = 2.54 \text{ cm}$$
$$100 \text{ cm} = 1 \text{ m}$$
$$1\,000 \text{ m} = 1 \text{ km}$$

Next, we string out the conversion factors to make the conversion first from inches to centimeters, then from centimeters to meters, and finally from meters to kilometers.

$$(6 \text{ in.})\left(\frac{2.54 \text{ cm}}{\text{in.}}\right)\left(\frac{\text{m}}{100 \text{ cm}}\right)\left(\frac{\text{km}}{1\,000 \text{ m}}\right)$$

Canceling out the common units in the numerators and denominators, we reduce the conversion to

$$\frac{(6)(2.54)}{(100)(1\,000)} \text{ km}$$

or

$$6 \text{ in.} = \underline{0.152\,4 \times 10^{-3} \text{ km}}$$

Since conversion factors are nondimensional and have an absolute value of 1, they can be divided as well as multiplied into the original units; that is, 100 cm/m has the same magnitude as 1 m/100 cm. The choice of which unit to place in the numerator or denominator depends on which unit is being eliminated in the conversion process.

Example 6.5

Convert 136 N to pounds force.

SOLUTION

Let us assume that we do not have the direct conversion factor between pounds force and newtons. Instead, let us work in more fundamental units relating force to mass, length, and time.

Given:

$$1 \text{ N} = 1 \text{ kg} \cdot \text{m}/\text{s}^2$$
$$454 \text{ g} = 1 \text{ lb}_m$$
$$1\ 000 \text{ g} = 1 \text{ kg}$$
$$2.54 \text{ cm} = 1 \text{ in.}$$
$$100 \text{ cm} = 1 \text{ m}$$
$$12 \text{ in.} = 1 \text{ ft}$$

$$1 \text{ lb}_f \cdot \text{s}^2/\text{ft} = 1 \text{ slug} = 32.2 \text{ lb}_m$$

Then, multiplying 136 N by the appropriate conversion factors, we get

$$\left(\frac{136 \text{ kg} \cdot \text{m}}{\text{s}^2}\right)\left(\frac{1\ 000 \text{ g}}{\text{kg}}\right)\left(\frac{\text{lb}_m}{454 \text{ g}}\right)\left(\frac{100 \text{ cm}}{\text{m}}\right)\left(\frac{\text{in.}}{2.54 \text{ cm}}\right)\left(\frac{\text{ft}}{12 \text{ in.}}\right)\left(\frac{\text{lb}_f \cdot \text{s}^2}{32.2 \text{ lb}_m \cdot \text{ft}}\right)$$

This can be reduced to the following by canceling the units appropriately:

$$\frac{(136)(1\ 000)(100)}{(454)(2.54)(12)(32.2)} \text{ lb}_f$$

Completing the multiplications and divisions, we get

$$136 \text{ N} = 30.6 \text{ lb}_f$$

In this last example, the string used factors first to convert the mass in kilograms to mass in pounds mass, then to convert the length in meters to length in feet. The last conversion was necessary to convert the mass in pounds mass to mass in slugs, which could then be redefined in units of pounds force, feet, and seconds squared, leaving the desired unit of pounds force.

In many engineering problems there are units hidden in the constants of equations known to be dimensionally consistent. These **hidden units** can pose a significant problem unless they are recognized before the conversion string is developed. Once the hidden units are identified, the conversion of units can take place as previously demonstrated.

Example 6.6

The heat flow q through a certain material of given thickness and cross-sectional area can be expressed by the dimensionally consistent relation

$$q = 5.453 \ \Delta T$$

where

q = heat flow in watts
ΔT = temperature difference in degrees Celsius

What are the hidden units in this equation?

SOLUTION

Given that the units of q are in watts and the units of ΔT are in degrees Celsius, then the hidden units must be in watts per degree Celsius and must be associated with the constant 5.453.

$$\text{W} = \frac{\text{W}}{\text{°C}} \cdot \text{°C} = \text{W}$$

It is fortunate that conversion tables and calculator conversion programs are readily available to students of engineering. Many direct conversions can now be made quickly in the calculator with little input to the conversion process that is actually taking place. Yet, there is also a danger in too much convenience, for it is important to understand the conversion process. We can never predict when these conversion conveniences may not be at our disposal. By memorizing the fundamental conversions listed in Box 6.1, you will be able to accomplish most conversions by appropriately multiplying and dividing the conversion factors as previously demonstrated.

Box 6.1

Fundamental Conversion Factors

Time	Mass	Length
60 s = 1 min	1 000 g = 1 kg	10 mm = 1 cm
60 min = 1 h	454 g = 1 lb_m	100 cm = 1 m
24 h = 1 d	1 slug = 32.2 lb_m	1 000 m = 1 km
365 d = 1 yr	1 slug = 1 $\text{lb}_\text{f} \cdot \text{s}^2 / \text{ft}$	2.54 cm = 1 in.
		12 in. = 1 ft
		3 ft = 1 yd
		5 280 ft = 1 mi

PROBLEMS

To review the subject matter in this chapter, take a piece of paper, answer the questions, fill in the blanks, and match terms.

6.1 Velocity is a fundamental dimension. True or false?

6.2 Give two examples of a dimensionless quantity.

6.3 The fundamental dimensions of pressure are F/L. True or false?

6.4 Match each term in the **Term** column with the appropriate units in the **Units** column:

Term	Unit(s)	Term	Unit(s)
Charge	cubit	Velocity	kilowatt-hour
Energy	watt	Pressure	coulomb
Power	km/s	Length	A
Current	Pa		

6.5 A pound mass is conceptually the same thing as a pound force. True or false?

6.6 Fill in the blank(s):

a. An erg is a unit of _____ in the _____ system of units.
b. One newton equals _____ dynes.
c. A watt is defined to be _____ .
d. One radian equals _____ degrees.
e. One centimeter equals _____ feet.
f. A pascal is a unit of _____ .

6.7 Which of the following are derived dimensions?

a. Energy
b. Length
c. Volume
d. Temperature
e. Luminous intensity
f. Area
g. Velocity
h. Time
i. Charge
j. Mass

6.8 List the fundamental or base dimensions in each of the unit systems that we have discussed.

6.9 What is the difference between absolute unit systems and gravitational unit systems? Which of the four unit systems that we have described are absolute (English, CGS, SI, engineering)?

6.10 What is the advantage of the engineering system of units? Why has this unit system caused so much confusion?

Work the following dimension and unit problems.

6.11 Find all the SI errors in the following statement, assuming that the intent is to use only SI units.
   The pressure and temperature of the gas entering the compressor are 120 kilo pascals and 290°K, respectively. The pressure ratio across the compressor is 6 to 1 and the temperature at the turbine inlet is 1100°C. On leaving the turbine the air enters the nozzle and expands to 100 kPa.

6.12 Where appropriate, correct the SI format of the following units:

a. 10.6 MN
b. newtons per meter^2
c. square seconds
d. 3123.6 × 10^3 sec
e. 1 J/s = 1 watt = 1 W
f. .301 kgs

6.13 Express the following quantities in terms of their fundamental dimensions using the "absolute" dimensions M, L, and t:

a. Force
b. Velocity
c. Power
d. Work
e. Pressure
f. Acceleration
g. Mass density
h. Volume
i. Torque

6.14 Which of the following are true?

a. 30°C > 32°F
b. 273 K > 460°R
c. −30°F > 0°C
d. 600 K > 400°C

6.15 Which of the following are false?

a. 454 g = 1 lb$_f$
b. 32.2 slug = 1 lb$_m$
c. 1 lb$_m$ = 1 lb$_f$
d. 1 000 g = 1 kg

6.16 Determine the fundamental dimensions of V in the following equation for a pendulum:

$$V = -mgL \cos \theta$$

where m is the pendulum mass, g is the acceleration of gravity, L is the pendulum length, and θ is the angular position of the pendulum.

6.17 In the following dimensionally homogeneous equation, P is a pressure; F is a force; y, r, and h are lengths; c is a reciprocal length; and α and β are angles. What are the fundamental dimensions of s?

$$P = \frac{Fy}{cs^4} \left(\frac{r}{h} \sin 2\alpha + \sin \beta \right)$$

6.18 The equation $A = (E - 2F) \sqrt{2BC/D}$ has terms that have the following fundamental dimensions:

$$B \sim M \cdot L^{-2} \cdot t^{-2} \qquad D \sim M \cdot L^{-3}$$
$$C \sim L \qquad E \sim L$$

If the equation is dimensionally homogeneous, what are the fundamental dimensions of A and F?

6.19 The thrust propelling a rocket is given by the equation

$$F = \dot{m}N \sqrt{kRT}$$

where R is a gas constant that for air is 53.35 ft · lb$_f$/(lb$_m$ · °R); k is a dimensionless ratio; T is the air temperature (°R); and N is a dimensionless mach number. What are the fundamental dimensions of $\dot{m}$ and what does $\dot{m}$ physically represent?

6.20 The transient response of an electric circuit having a resistor, inductor, and capacitor in series is a function of sin ωt, where t is time and

$$\omega = \sqrt{\frac{1}{LC} - \frac{R^2}{4L^2}}$$

Given that R is the resistance and L and C are the inductance and capacitance, respectively, what are the fundamental dimensions of ω? What does ω physically represent?

6.21 The drag on an object in an airstream can be expressed by the following equation, where D is the drag force, ρ is the mass density of air, A is the cross-sectional area of the object, V is the speed of air relative to the object, and C_D is the drag coefficient. What are the fundamental dimensions of C_D?

$$D = \frac{1}{2}\, C_D \rho A V^2$$

6.22 In the following equation, how do the fundamental dimensions of C relate to the fundamental dimensions of B?

$$P = \mu g H \sqrt{B + C^2}$$

6.23 Use the fundamental conversion factors and temperature relationships to convert the following:

a. 1.53×10^4 mph to km/s
b. $-73.7°C$ to °R
c. 786.72 N to dynes
d. 14.1 psi to MPa
e. 30.1 rad/s to rpm
f. 10 hp to kW (see Appendix A for hp conversion factors)
g. $-410.1°F$ to K
h. 3.12×10^4 N to lb_f
i. 0.15 $lb_m/in.^3$ to kg/m^3
j. 0.042 $kg \cdot m^2$ to $lb_m \cdot in.^2$
k. 987 slug to g
l. 10^3 erg to $ft \cdot lb_f$

6.24 In Newton's law of gravitation ($F = Gm_1m_2/r^2$), the gravitational constant G has the value of 6.67×10^{-8} $cm^3/(g \cdot s^2)$. By using conversion factors, determine the value of G in SI and in the English and engineering systems of units.

6.25 Determine the weight in newtons of water contained in a water tower having a cylindrical water tank of height 7.5 m and diameter 10 m. Assume that the water tank is three quarters full.

6.26 The Fourier heat conduction equation can be written as

$$q = \frac{kA(T_2 - T_1)}{x}$$

where q is the heat transferred per second, T_1 and T_2 are the temperatures across the heat-conducting element, k is the thermal conductivity, x is the element width, and A is the area through which heat is being transferred. In this equation x/k is sometimes called the R *factor* and has units of $h \cdot ft^2 \cdot °F/Btu$. What is the rate of heat being transferred through an external wall of area 150 ft^2 if the wall has an inside temperature of 75°F, an outside temperature of 15°F, and an effective R value of R11? At an electric energy charge of 10¢ per kilowatt-hour, what would it cost to keep this wall at 75°F for 8 hours?

6.27 A high-power line is strung between two base towers 75 m apart. If the power line has a mass per length of 0.5 kg/m, determine the weight of the cable (neglect the power line sag).

6.28 How many Btu's of heat are given off from a 100 W lightbulb that is on for 5 hours?

6.29 An electric motor is being used to lift a 50 kg mass at a rate of 0.2 m/s. If the efficiency of this process is 85 percent, what power is being used by the electric motor in kilowatts and horsepower?

6.30 Bernoulli's equation for an inviscid (frictionless), incompressible fluid flowing through a pipe can be written in the form

$$\frac{p}{\gamma} + \frac{v^2}{2g} + Z = \text{constant}$$

where γ is the specific weight (F/L^3) and at some cross section in the pipe, p is the pressure, v is the average flow velocity, and Z is the height of the pipe above some reference point. Given a pipe of constant cross section in which water is flowing steadily ($v = $ constant), what must be the difference in heights between two points in the pipe to cause a pressure difference of 0.12 MPa?

6.31 Ohm's law states that the voltage drop E across a resistor of resistance R is related to the current I through the resistor by the equation $E = IR$. In addition, the power P dissipated in the resistor is related to I and E by the equation $P = IE$. Given a voltage drop across R of 10 V, determine the power in kilowatts being expended in a 100 Ω resistor.

6.32 The tip deflection δ of a uniformly shaped cantilever beam that is under a distributed load w can be expressed by the equation

$$\delta = \frac{Wl^3}{8EI}$$

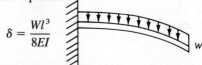

where

W = total weight of beam = wl
w = weight per length of beam
l = length of beam
E = modulus of elasticity
I = area moment of inertia

What is the deflection δ in inches for a steel beam having $E = 30 \times 10^6$ lb$_f$/in.^{2}, $I = 0.33$ in.^{4}, $w = 15$ lb$_f$/ft, and $l = 15$ ft?

6.33 Experiment shows that the mass of an electron varies with speed, changing appreciably as the velocity approaches the speed of light. Given that m_o, the "rest mass" of an electron in the following equation, is 9×10^{-28} g and the velocity of light c is 3×10^{10} cm/s, what would the mass m of the electron be if it were traveling at 80 percent of c?

$$m = \frac{m_o}{\sqrt{1 - \left(\dfrac{v}{c}\right)^2}}$$

6.34 If the thrust propelling a rocket is given by the equation of problem 6.19, what is the thrust in pounds force if $T = 530°$R, $N = 2.1$, $k = 1.4$, $\dot{m} = 10$ lb$_m$/s, and $R = 53.35$ ft·lb$_f$/(lb$_m$·°R)?

6.35 Thermal radiation between two surfaces may be described by the Stefan-Boltzmann law

$$q = C(T_2{}^4 - T_1{}^4)$$

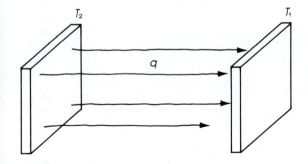

where T_2 and T_1 are the *absolute* temperatures of the emitting and receiving surfaces, respectively, C is a constant that depends on the shape, area, and surface characteristics of the body that is emitting radiation, and q is the rate of thermal energy being transferred per area of radiation-emitting surface. What is the power being radiated between two parallel surfaces 100 m^2 in area if $C = 0.2$ J/(m^2·s·K^4), $T_1 = 30°$C, and $T_2 = 1\,000°$C?

6.36 It is desired to measure the unknown resistance of the electrical "black box" shown here. We connect a 10 V battery and an ammeter (device that measures the current through it and has negligible resistance to current flow through it) to complete the circuit as shown and note a 0.1 A reading on the ammeter scale. What is the resistance of the black box in ohms?

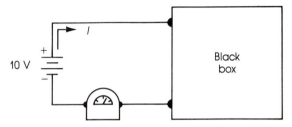

6.37 If the average density of the earth's atmosphere up to an altitude of 1 mile is 2.13×10^{-3} slug/ft^3, what is the total mass in slugs, pounds mass, and kilograms of that portion of the atmosphere?

6.38 The Sommerfeld number is an important number in journal-bearing lubrication and is related to the viscosity μ, the load per projected bearing area P, the shaft speed n in revolutions per second, the journal radius r, and the minimum lubricant thickness c by the equation

$$S = \left(\frac{r}{c}\right)^2 \frac{\mu n}{P}$$

If $\mu = 4 \mu$reyns (pronounced "microreyns") and a reyn is a lb$_f$·s/in.^{2}, $r = 2$ in., $c = 0.001\,5$ in., $P = 222$ psi, and $n = 30$ rev/s, what is the Sommerfeld number and what are its units?

6.39 The earth has an approximate mass of 6×10^{24} kg and orbits the sun in a near-circular path of diameter approximately 1.5×10^8 km. Assuming that the mass of the sun is approximately 2×10^{30} kg, apply Newton's law of gravitation to find the force of attraction between the earth and the sun in newtons. Now approximate the orbital speed of the earth in kilometers per hour by applying the centrifugal force equation

$$F_e = \frac{m_e V^2}{R}$$

where F_e is the force of the sun's attraction on the earth, m_e is the mass of the earth, R is the orbital radius, and V is the orbital speed. Assuming that the earth's orbit is perfectly circular, calculate the time to orbit the sun in days, using V.

7 The Technical Library

CONTENTS

The Technical Library
Reference Works
Standards and Specifications
Government Documents
Technical Reports
Two Reference Searches
Patents
The Drama of Patents

Throughout their careers, engineers and technologists face the formidable challenge of remaining technically competent. A thorough familiarity with the contents and use of the technical library will help.

THE TECHNICAL LIBRARY*

Your future engineering career will include finding information and data as well as applying it. When you begin your career as a practicing engineer, you make not have all the information sources discussed in the following sections at your disposal. Probably you will have a smaller library available to you that will include the most used reference tools for your firm and that may or may not be under the direction of a librarian. You and the library staff may have access to a large academic or public library with a special science and business collection to supplement your basic collections. This section will concentrate on the kind of information tools available at these larger libraries. This should help you handle those college assignments that are training you to be an engineer as well as familiarize you with the information tools that you will rely on as you pursue your profession.

Types of Technical Libraries

Technical libraries vary from university to university. For example, some of the larger campuses have special science or engineering "only" libraries where most of the technical reference materials are located. If you are studying to be a chemical engineer on a campus with this kind of library system, you will periodically need books and information from the chemistry library; electrical engineers may need some material from the physics library; civil engineers may want maps and information from the geology library.

Other campuses have a separate scientific and technical library where most of the sciences and engineering reference materials are collected together. Business information and U.S. government documents may be in another

---

*Contributed by Barbara Hedges, Paula Higgins, and Julia Rholes, reference librarians, Texas A&M University.

location. Finally, many universities that concentrate on science and technology have a centralized library system where all the reference materials are located.

In addition to knowing the general kind of material available in these libraries, it is extremely helpful for library users to familiarize themselves with the layout and organization of the library or libraries they will be using. Many libraries offer general orientation tours at the beginning of each school year or semester. In addition, there may be specially prepared written guides for the library. Some libraries offer specific courses to aid in using their facility; or library instruction may be included in courses such as English or technical writing. Some library staffs offer special tours or classroom lectures to accompany a class assignment in using the library. Check with your library and your classroom instructor to find out what is available to help you. Needless to say, it is prudent to learn something about your library facilities before that last-minute assignment is due.

Library Organization

All libraries are divided into basic sections; however, individual libraries have different ways of managing those sections. Some of the basic sections are described here.

CIRCULATION   In this section one checks out the books that may be removed temporarily from the library. Special materials, such as maps, reserve materials, microforms, government documents, or films, may circulate from their own departments. Usually, to check out materials, student identification or library identification cards are needed.

CARD CATALOG   The card catalog is usually located near the entrance to the library building. It contains many rows of cabinets with drawers of cards. The cards are used to locate books for which either the author or the title is known. The card catalog can also be used to locate books on a given subject, but it cannot locate chapters in books on a given subject, nor can it locate periodical articles on a given subject. To locate articles on a given subject, one uses a periodical index. The periodical index will be examined in more detail in a later section.

Cards are filed alphabetically in the card catalog. For each book there will be a card that begins with the book's title, followed by additional information about the book's contents. There will also be a card for each author of each book. This enables you to look up either the author or the title alphabetically and determine if the library has the book. Subject cards, which identify books pertaining to particular subjects, are either segregated from the other cards or integrated alphabetically within the author and title cards. Larger libraries usually have a separate card catalog organized by subject. The cards will note the publisher and publishing date; the book's call number appears in the upper left-hand corner.

Shown here is a catalog card for the *McGraw-Hill Dictionary of Scientific and Technical Terms.* The call number is given in the upper left-hand corner. (Courtesy of the author)

A call number is assigned to each book to arrange books in the library. Call numbers are composed of letters and numbers so that books may be shelved alphabetically and numerically in the order of the call number. The same call number you find on the catalog card will be listed on the book. Call numbers are assigned according to the Library of Congress System or the Dewey Decimal System; both of these systems group books and periodicals on the same subject together.

The Dewey Decimal System assigns numbers by subject. Technology is 600–699; engineering is 621, and electrical engineering is assigned the sub-number 621.35. Additional letters and numbers follow this in the call number.

The Library of Congress system uses a combination of letters and numbers to indicate a subject. The letters T–TZ are the first part of the call number for any engineering book. For example, call numbers for electrical engineering books begin with the letters TK, while the letters assigned to chemical engineering are TP.

JOURNALS AND JOURNAL HOLDING LISTS The more current information in a technical field like engineering is found in the technical journals. Some libraries, usually the smaller ones such as a branch engineering library, will shelve their journals alphabetically in a separate section. Larger libraries, which subscribe to many journals, will shelve them by a call number, just like

A computer-output microfiche journals list (right) is pictured with a microfiche reader. (Courtesy of the author)

the call numbers that libraries use for books. Engineering journals will utilize call numbers very close to the call numbers assigned to engineering books.

A title card for a journal may be filed in the card catalog, sometimes followed by a card that indicates which years and volumes of a journal the library owns. Often there is a separate list for journal holdings. When there is a separate list, such as a microfiche list or computer printout, the journals the library subscribes to will be listed alphabetically. In addition to the title, the journal list will include the call number for those journals shelved by call number, and what years and volumes are available.

RESERVE COLLECTION   Many libraries will have a section or separate room where professors may place materials for students in their classes to read. This may include books from the library collection, including reference books, occasionally the professor's own books, photocopies of paper of journal articles, and photocopies of sample test questions or answers to homework problems. There is usually a much shorter check-out period for reserve materials.

INTERLIBRARY SERVICES   You may be able to obtain photocopies of articles from journals your library doesn't own, or a short-term loan of a book your library doesn't own, through the Interlibrary Services Department. The use of these special services may be limited to faculty and graduate students at some university libraries. Your library will request another library, which has the material needed, to send a volume or copy. You will usually be expected to pay the fee charged by the other library, which would include photocopying costs.

OTHER SPECIAL DEPARTMENTS   Many libraries will have special sections for maps, U.S. government documents, and technical reports. Sometimes the map section may be located in a geology branch library; technical reports will often be found in an engineering branch library or with U.S. government documents. There are many possible arrangements of these sections.

REFERENCE SECTION   The reference section of the library will contain some of the important books you will need in your engineering career. Reference books include dictionaries, encyclopedias, handbooks, directories, periodicals indexes, and bibliographic guides to technical literature. These books contain short discussions of a word or subject or small pieces of information or data. They also serve as important guides by citing books and journal articles where more information about a subject can be found. You can find the dimensions of steel beams, the boiling point of methanol, or citations to journal articles on the hydrocracking procedures used in geothermal wells.

REFERENCE WORKS

There are special dictionaries, encyclopedias, handbooks, catalogs, directories, and periodical indexes for engineering as well as for sciences, such as chemistry or physics, that have engineering applications. In the following sections the different kinds of reference books will be discussed.

Technical Dictionaries

Technical dictionaries may include illustrations and formulas as well as the definitions of words. The definitions tend to be more detailed and precise than in general dictionaries because of the precise way scientists and engineers use language in their theory and applications. There are a multitude of dictionaries for science and engineering. The following dictionary is one example of what is available; others are listed in Appendix B.

> *McGraw-Hill Dictionary of Scientific and Technical Terms.* 2nd ed. Edited by Daniel N. Lapedes. New York: McGraw-Hill, 1978.

This dictionary contains entries for all areas of science and engineering. A useful feature is that, along with the definition of a term, the dictionary gives the discipline, or subject areas, in which the definition of the word is used. This is particularly helpful when a word has more than one meaning.

Bibliographies, Literature Guides, and Book Catalogs

Bibliographies, literature guides, and book catalogs are closely related types of reference sources. A *bibliography* is a listing of sources relevant to a particular topic. The sources can be books, technical reports, patents, or journal articles. You can locate books in the Library of Congress System by using any subject heading with the subheading "Bibliographies," for example, "Chemical Engineering—Bibliographies." *Literature guides* are bibliographies, but they also contain suggestions for use of the sources listed. *Book catalogs* are the catalogs of libraries published in book form. Some special libraries may have journal articles and patents in their catalogs as well as books. They may also have special subject headings on a special subject classification that may make them easier to use for specific technical subjects.

Following are a few bibliographies, literature guides, and book catalogs helpful to engineers:

> Mount, Ellis. *Guide to Basic Information Sources in Engineering*. New York: Wiley, 1976. A detailed guide to the use of engineering libraries and engineering literature.

> Chen, Ching-Chih. *Scientific and Technical Information Sources*. Cambridge, Mass.: MIT Press, 1977. This work was written as a reference guide for science and engineering libraries, but is useful for anyone interested in science and engineering.

> American Society for Engineering Education, Engineering Schools Library Division. *Guides to the Literature*. Washington, D.C.: American Society for Engineering Education, 1970. Sixteen brief guides to engineering literature including chemical engineering, computers, electrical and electronics engineering, environmental sciences, industrial engineering, medical engineering, metals and metallurgical engineering, and transportation engineering.

> *Engineering Societies Library (New York) Classed Subject Catalog*. 13 vols. Boston: G. K. Hall, 1963. With annual supplements from 1964 to the present. These volumes include the 185 000-volume collection of books, journals, technical reports, and other material in all fields of engineering. The yearly supplement lists the most recent material.

> *Northwestern University Transportation Center Catalog*. 12 vols. Boston: G. K. Hall. A catalog of books and serials on the subject of transportation engineering and management.

Technical Encyclopedias

In general, a technical encyclopedia is written for the person who is beginning to learn about a technical field. An expert in one area might use a technical encyclopedia to find out basic information in another area of engineering or science. Technical language is used in technical encyclopedia articles; however, technical encyclopedias are usually not as detailed and comprehensive as a book or a journal article written for the expert in a field. The articles in a technical encyclopedia provide an introduction and overview of various technical subjects. The articles will often include a bibliography for the reader who wishes to know more about a subject. A selection of a few major technical encyclopedias can be found in Appendix B.

Handbooks

Handbooks are the key reference tools in engineering. Because they provide a comprehensive review of a topic for the expert, they lack the introductory technical discussions and explanations found in technical dictionaries, encyclopedias, and textbooks. Usually they are written and compiled by experts. Handbooks contain numerical data as well as formulas and technical discussions for the field they cover. A few of the many handbooks in different areas of engineering are listed and discussed in Appendix B.

Periodical Indexes and Abstracts

To find articles about a particular subject in technical journals or in proceedings or meetings, one uses periodical indexes. Periodical indexes are subject-oriented bibliographies of journal literature published during a certain time period. The index will enable the researcher to find a citation to an article, including the title of the article, the name of the author, the name and volume of the journal, and the pages where the article is printed. An abstracting service will also have abstracts, or short paragraph summaries, about the article.

There are at least 2 500 different indexing and abstracting services for science and technology. Most of them are primarily index journal articles, but some will include a few books, technical reports, or patents. A listing and discussion of a few of the major indexes that are useful for engineering subjects can also be found in Appendix B.

On-Line Bibliographic Data Bases

Most printed indexes to journal or magazine articles possess computerized counterparts called data bases. Trained computer analysts in university and company libraries search the many data bases on topics of interest to the user. A computer literature search is a quicker and more efficient method of obtaining information. Instead of someone thumbing through volume after volume of paper copy, as in a manual search, the computer displays the results almost instantly from records covering a number of years. Since the majority of data bases were begun around 1970, the older citations to journal articles are not available through a computer search. Printed indexes must be used instead.

The computer searches for any word requested by the user in the titles, descriptors, and abstracts, if available, of the records in the data bases. For example, if someone were interested in the effects of fatigue on turbogenerators, the word "fatigue" would be combined with the phrases "turbogenerators" or "turbine generators" and entered into the computer via a terminal.

The *Engineering Index,* pictured here, is a major comprehensive abstracting service for locating references to articles on engineering topics. (Courtesy of the author)

Three commercial vendors who provide computer access to these data bases are Lockheed Missile Corporation, System Development Corporation (SDC), and Bibliographic Retrieval Services (BRS). Each has a different list of available data bases, and each uses different computer programs. Some data bases are unique to a particular vendor, and some are provided by all three. Since these companies are in the business of providing computer indexes to make money, the user is charged a fee for each search.

There are many data bases relevant to the engineering profession. Compendex, which began in 1974, is the on-line version of *Engineering Index,* one of the most comprehensive indexes in this field. It covers journal articles, conference proceedings, publications of engineering societies, and books. The data base containing the National Technical Information Service (NTIS) *Government Reports Announcements* indexes the technical report literature written since 1964 in a broad range of areas, including engineering and technology. A third general coverage data base is derived from the *Science Citation Index,* which indexes journal articles published since 1974.

Some data bases cover only specific fields of engineering. Information Services in Mechanical Engineering (ISMEC) is the on-line version of *ISMEC Bulletin* and indexes more than 250 journal titles as well as books, reports, and conference proceedings. Another specialized data base is International Information Services in Physics, Electrotechnology, Computers, and Control (INSPEC), which corresponds to *Physics Abstracts, Electrical and Electronic*

Abstracts, and *Computer and Control Abstracts.* It covers journal articles, conference proceedings, reports, patents, and books published since 1969.

Transportation Research Information Service (TRIS) covers literature in the areas of air, highway, maritime, railway, and urban transportation. It indexes journal articles, technical reports, project résumés, conference proceedings, and dissertations. The *Society of Automotive Engineers* (SAE) *Abstracts* data base, available since 1965 from System Development Corporation, covers the SAE conference papers concerning self-propelled vehicles.

Other data bases cover the fields of petroleum and energy engineering, industrial engineering, and chemical engineering. Some data bases index a particular kind of document such as patents, dissertations, government publications, books, or federal regulations. Data bases are proliferating rapidly, and the future looks bright for improved access to vital information.

Manufacturers' Catalogs and Directories of Manufacturers

CATALOGS Manufacturers' or trade catalogs are an important source of information for engineering students and for practicing engineers. The catalogs contain pictures, descriptions, diagrams, schematics, specifications, dimensions, instructions on use, and performance test standards of the product. In addition, the catalogs will contain addresses of the regional and local sales offices of the company. However, in most cases, prices of the products are not provided in these reproductions of manufacturers' catalogs. The few prices given are not to be used as quotations, since participating manufacturers reserve the right to raise prices without notice. Students find such information invaluable in working on design problems. In practice, extensive research and development of a component will not be required if the necessary part is available for purchase elsewhere. Workable and marketable ideas on designs can be stimulated by reviewing catalogs displaying similar products. Also, important information on smaller units can be extracted for use in larger design projects.

Unfortunately, these catalogs are sometimes difficult to obtain. The *Thomas Register of American Manufacturers* and *MacRae's Blue Book* can be used to identify companies that manufacture a certain product. The engineer can then request a catalog from the company. However, since designers usually need this information immediately, this is a waste of valuable time. Some libraries laboriously compile a private file of trade catalogs and then must provide some kind of index to the collection. On the other hand, some of the collections of trade catalogs available from commercial publishers include:

> *Thomas Register Catalog File (ThomCat).* New York: Thomas Publishing, 1905 to the present. Annual. Multivolume set that contains more than 800 catalogs bound in alphabetical order by company name. It covers all areas of manufacturing.

A trade catalog for semiconductors provides information on products and manufacturers. (Courtesy of the author)

Sweet's Catalog File. New York: McGraw-Hill Information Systems Company, 1914 to the present. Annual. This contains trade catalogs of companies in the areas of architecture, plant engineering, and machine tools.

EEM; Electronics Engineers Master Catalog. Garden City, N.Y.: Unit Technical Publications, 1958 to the present. Annual. This two-volume set incorporates a product index and directory, a manufacturer's and sales office directory, a trademark directory, and manufacturer's catalog pages in the area of electronic and electromechanical components, systems, and equipment.

Chemical Engineering Catalog. New York: Reinhold Publishing, 1916 to the present. Annual. This collection includes an equipment and chemical materials index, a trade name index, a sales office directory, and trade catalog pages for process equipment and plant design and construction.

VSMF (Visual Search Microfilm). Englewood, Colo.: Information Handling Services. Monthly. This service provides thousands of trade catalogs on microfilm cartridges, frequently updated, along with indexes to find the correct frame of microfilm. One of their systems is arranged by product with an index to manufacturers, another is organized by vendor name with an index by type of product. The catalogs cover many different areas of manufacturing. It also provides microfilm of standards and specifications.

DIRECTORIES Many directories of manufacturers will contain product information from the manufacturer's catalogs, as well as the addresses of companies producing specific types of products. Examples of these directories for several industries follow:

Solar Engineering Master Catalog and Solar Industry Index. Dallas: Solar Engineering Publishers, 1979 to the present. Annual.

Electronic Design's Gold Book, Master Catalog and Directory of Suppliers to Electronics Manufacturers. Rochelle Park, N.J.: Hayden Publishing, 1974/75 to the present. Annual.

Transistor D.A.T.A. Book. Pine Brook, N.J.: Derivation and Tabulation Associates. Annual.

Keystone Coal Industry Manual. New York: McGraw-Hill Mining Publications, Mining Informational Services, 1969 to the present. Annual.

Soap/Cosmetics/Chemical Specialties Bluebook. Issued as a buyer's guide annually for the periodical *Soap/Cosmetics/Chemical Specialties*.

Chem. Sources. Flemington, N.Y.: Directories, 1958 to the present. Annual.

STANDARDS AND SPECIFICATIONS

Standards and specifications are detailed instructions for designing and manufacturing a product. They offer definitions, required performance tests, calibration curves, accepted dimensions or composition of the product, and the underlying principles behind the methods of operation of the product. Standards improve the quality of the product and help to simplify modern life. Parts are easily replaced, since each type follows a similar manufacturing procedure.

Standards are issued by several different agencies on just about everything. There are standards for seat belts, spark plugs, and rearview mirrors; for chemicals and plastics; for structural steel, cement, and brass wire—to mention only a few. The federal government is one of the most prolific publishers of standards and specifications (Figure 7.1). Military specifications for all branches of the armed forces are issued by the Department of Defense and accessed through its *Index of Specifications and Standards*. The General Services Administration prepares standards for common items used throughout the federal government that are accessed through the *Index of Federal Specifications and Standards*. Nongovernmental organizations also issue standards. These include engineering societies such as the Society of Automotive Engineers (SAE) and the American Society of Mechanical Engineers (ASME); national organizations such as the American National Standards Institute (ANSI) and the American Society for Testing and Materials (ASTM); and trade associations such as the National Electrical Manufacturers Association (NEMA) and the National Fire Protection Association (NFPA). Standards are also issued by foreign agencies such as the British Standards Institute (BSI) and international bodies such as the International Organization for Standardization (ISO) and the International Electrotechnical Commission (IEC).

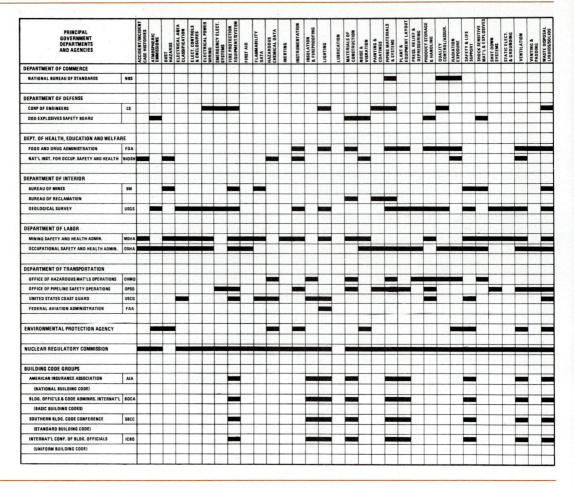

FIGURE 7.1

Government sources for published design codes, standards, recommended practices, and regulations for plant facilities and ancillaries. (Courtesy of C. R. Burklin, Brown and Root, Inc.)

GOVERNMENT DOCUMENTS

The U.S. government publishes an enormous number of documents that can be useful to engineers, including a wide variety of technical bibliographies, dictionaries, handbooks, directories, and thesauri. It also compiles and publishes statistical information on topics ranging from the number of tons of

A microfiche file of NTIS (National Technical Information Service) reports is shown here. They often are located in the government documents department. (Courtesy of the author)

steel produced in a given year to the average starting salary of a mechanical engineer. All of these sources of information can be helpful. There are, however, two special types of government documents that are extremely important to engineers: technical reports and patents. Because of their importance, they will be discussed in greater detail later.

Unfortunately, many engineers do not use government documents as much as they might. One reason for this is that it is not always easy to find information in government documents. Most libraries keep their collection of government documents separate from their regular book collection by listing them in a separate book catalog, *The Monthly Catalog*. However, even *The Monthly Catalog* does not cover all government documents. Therefore, sometimes several different government or commercially published indexes must be searched to find the needed information. Some of these indexes are well organized and easy to use; some are not. Thus, for some searches of government documents, the researcher will need the guidance of a librarian. But regardless of the professional assistance available, it is important to understand the nature of the government documents that are available and the basic procedure for finding them in order to successfully tap this important source of engineering information.

TECHNICAL REPORTS

A technical report documents the results or progress of a research or development program or project. Most technical reports are part of a numbered series. While these series usually do not have a title, each report will have its own serial code and numbers. These numbers and codes are usually of great importance in trying to locate technical reports.

Technical reports are particularly helpful because they are often more timely than periodical articles and because they can be more specific. Because of high publication costs, journal articles are usually limited in length; a technical report, however, can vary in length from a few pages to several volumes. The federal government is by far the largest publisher of technical reports, but corporations and associations also issue them. Although the majority of technical reports are available to the public, some are subject to security regulations, either military or commercial.

Technical reports may take several different forms. The most common type is the contract progress report. This report is written for the sponsor of the research, but is often available to the public. Less common but usually more valuable is the final report of contracted research. Another type of technical report is an individual author's preprint. Preprint reports often end up in a journal six months to a year later. There is also the institution report that tries to justify a budget or a corporate proposal report that is aimed at prospective customers. Finally, many technical reviews or state-of-the-art books are in technical report form.

Although most of the major periodical indexes such as *Chemical Abstracts, Engineering Index,* and *Metals Abstracts* will cover technical reports, there are a number of indexes specifically directed to technical report literature. These indexes provide a more comprehensive coverage of technical reports. They contain not only the standard author and subject indexes but also index by agency, report number, and contract number, which is extremely helpful information.

Most technical report indexes are published by the government; a few, however, are published privately, and these are usually more difficult to locate. To further complicate the matter, some private technical reports will not be covered by any index. Fortunately, some of these reports will be cataloged and can be found in the card catalog in the usual manner. It is important to note that some libraries maintain their own technical report indexes; these should be checked for hard-to-find technical reports. Some of the more important technical report indexes are found in Appendix B.

In most cases the government sells its technical reports in either a paper or microfiche format. Most libraries find it too expensive, both in terms of monetary cost and shelf space, to maintain paper copies of technical reports. Therefore, a library's technical report collection will often be on microfiche.

The physical location of technical report collections may vary from library to library. Since the majority of technical reports are published by the government, technical reports may be found in the government documents department of a library. They may also be kept in a science reference division because of their technical nature. Finally, technical reports sometimes may be found in a special microform department. Indexes to technical reports will usually be found near the actual reports.

TWO REFERENCE SEARCHES

The Design of Smoke Detectors

Suppose your engineering instructor had just given you the assignment to make a class presentation on the design features of smoke detectors. Because you are unfamiliar with smoke detectors, you are faced with a library search for pertinent information. First, you pose a set of questions. What is the theory behind their operation? What kinds of detectors are available? What do they look like? What light source do they use? What sort of sensor do they have? What is the density of the smoke they should detect? What is the size of the area their detection covers? What are their dimensions? What are the optimal operating conditions? Are circuitry diagrams available? What are their wiring and current power consumption requirements? What calibration curves are available? What kind of power supply is needed? What performance and maintenance tests must they pass? What ventilation do they require?

All these questions and others can be answered by perusing manufacturer's catalogs and standards issued by various organizations. For example, consider the problem of determining how smoke detectors sense smoke. If the Visual Search Microfilm (VSMF) system is available in your library, its use will facilitate the search, since it provides the standards and catalogs on microfilm. The *Product/Subject Master Index,* which is the starting point for using VSMF, provides the following information when checked under smoke detectors:

	VSMF Locator Code
Smoke (adj.)	
Detector,	
Annunciating/Switch Output	A-27-06
Broken Filter Bag	A-27-06
Indicating/Controlling/	
Recording/Metering,	
Source Emission Analysis	W-24-33
Detector Circuit	A-27-31

The VSMF locator codes dealing with smoke detectors (A-27-06, W-24-33,

A librarian uses Visual Search Microfilm (VSMF) to find information from manufacturer's catalogs and industry standards and specifications. (Courtesy of the author)

and A-27-31) are used throughout the system in the various locator indexes to obtain appropriate cartridge and frame numbers for the standards and manufacturer's catalogs.

FINDING STANDARDS   To find the standards on smoke detectors, look under the location codes in the *Industry Standard Locator Index*. The last two locator codes (W-24-33 and A-27-31) are not applicable. W-24-33 lists standards on analyzing smoke, and A-27-31 lists no standards. However, A-27-06 lists standards on "switches/annunciators, sensing smoke or combustible/toxic gas." Your library may not own all the standards listed. Check under as many standards as possible, since each may contain pertinent information.

One standard that explains how smoke detectors sense smoke is from the National Electrical Manufacturers Association (NEMA). The locator index shows the acronym of the standard, the standard number, the title of the standard, and the cartridge and frame number for finding the standard. In this case the information needed would be NEMA, SB 9-77, smoke detectors, 4007-0843 (Figure 7.2). Note that the second number changes periodically as entries are inserted or deleted.

Turning to the microfilm of the *Industrial Standards,* find cartridge number 4007, insert it into the microfilm reader, then find frame number 0843. This will be the first page of the NEMA standard number SB 9-77. The standard defines the various types of smoke detectors (cloud chamber smoke detector, ionization spot type smoke detector, photoelectric beam type detector, photo-

```
AAAAAAA              2222222   7777777
AAAAAAAAA            222222222  777777777   SWITCHES/ANNUNCIATORS,
AA    AA             22    22        77     SENSING
AA    AA    XXXXXX          22      77
AAAAAAAAA   XXXXXX          22      77
AA    AA                    22      77
AA    AA             22222222       77
AA    AA            222222222       77
```

SOCIETY	DOC NUM	DOCUMENT DESCRIPTION	CART/FRAME
**** 06 SMOKE; COMBUSTIBLE/TOXIC GAS		(CONT)	
BSI	BS 5446:-77 PT 1:	SPECIFICATION FOR COMPONENTS OF AUTOMATIC FIRE ALARM SYSTEMS FOR RESIDENTIAL PREMISES PART 1: POINT-TYPE SMOKE DETECTORS	R41-0692
JIS.	M 7603-62	SAFETY LAMP TYPE GAS TESTERS	J102-3415
JIS.	M 7625-77	INTERFEROMETER TYPE GAS ALARMS	Y -
JIS.	M 7626-65	HOT WIRE TYPE GAS ALARMS	Y -
NEMA	SB 9-77	SMOKE DETECTORS	4007-0843
NFPA	*NFPA 72C P-75	REMOTE STATION PROTECTIVE SIGNALING SYSTEMS (ANSI/NFPA 72C P-76)	4701-2118
SAE	AS 446-62	CARGO COMPARTMENT FIRE DETECTION INSTRUMENTS (TURBINE POWERED SUBSONIC AIRCRAFT)	1107-3354
U.L.	DRAFT 1484	PROPOSED FIRST EDITION OF THE STANDARD FOR RESIDENTIAL GAS DETECTORS AUGUST 1979	8014-0742
U.L.	SUBJECT 1484	PROPOSED FIRST EDITION OF THE STANDARD FOR RESIDENTIAL GAS DETECTORS; PROPOSED EFFECTIVE DATE AUGUST 15, 1979	8014-0741
U.L.	UL 217-80	SINGLE AND MULTIPLE STATION SMOKE DETECTORS FEBRUARY 20, 1980	8115-0224

* INDICATES ANSI APPROVED STANDARD
Y INDICATES NO ENGLISH TRANSLATION AVAILABLE

FIGURE 7.2

Page illustrating type of information available in a standard like the *Industrial Standards.* (Excerpted and reproduced from standard number SB 9-77 by permission of the National Electrical Manufacturer's Association)

electric spot type detector, and sampling smoke detector) and explains how they work. It also contains diagrams illustrating their operation. There were no military specifications listed under those location codes in the *Military Specification's Locator Index.*

SOME TYPICAL DESIGNS To obtain an idea of how a real smoke detector is designed, the manufacturers' catalogs need to be examined. Using the locator code A-27-06 in the *Design Engineering Locator Index* provides cartridge and frame numbers for pages from different trade catalogs. For example, the pages of Faraday Incorporated's *Catalog* on smoke detectors are in cartridge number 1124 starting with frame number 2639. These pages explain the operational principles of their smoke detectors and illustrate the various parts of it with detailed diagrams. If additional examples are required, the much larger *Master Catalog* can be checked for other trade catalogs on smoke detectors. The *Design Engineering* file has all its pages on smoke detectors together, which facilitates its use, but it contains a smaller number of catalogs.

The *Master Catalog* has its own *Index to Vendor Product Data: Product/ Service Listing*. Under"Smoke detector, ambient," this index gives a "see reference" to "Switch, sensing smoke/combustible gas." This indicates that the names of the manufactureres who produce smoke detectors will be listed under the second heading. In addition, twenty-nine manufacturers' catalogs not listed in the *Design Engineering* file are listed in the *Master Catalog*. The cartridge and frame numbers are given to find the correct catalog in the *Master Catalog* microfilm. However, the frame number indicated is the first page of the entire catalog for one particular manufacturer. The catalog must be searched for the pages on smoke detectors.

ALTERNATIVE SOURCES If the VSMF source is not available in your college or company library, the same information can be obtained by other methods. A review in the *Thomas Register of American Manufacturers* under fire alarm systems will disclose companies that make this product. Checking the *Thomas Register Catalog File* for catalogs of these companies provides the same information given in the *Design Engineering* file and the *Master Catalog*. The *ThomCat*, however, is a much smaller collection of trade catalogs. The American Society for Testing and Materials and the American National Standards Institite both provide annual indexes that can easily be checked for pertinent standards. Trade association standards such as those of the National Electrical Manufacturers Association are much more difficult to come by.

Dust Removal and Disposal

The volcanic eruption of Mount St. Helens caused a great deal of destruction. It also left a huge cleanup problem behind. There were two aspects to the cleanup problem: dust removal and disposal or recycling of the volcanic dust. Imagine you are an environmental engineer given the responsibility for the cleanup problem. Where could you go to find out what related work has been done in this area?

To find information on these problems, first consult the *LC List of Subject Headings,* then the card catalog. Subject headings of potential use might include "Volcanic ash, tuff," "Volcanological research," or "Lava." These headings would lead you to books listed in the card catalog specifically about volcanoes. One recent book is *Volcanology,* by Howell Williams and Alexander R. McBirney (1979). However, since this is a new problem in the United States, there may not be enough relevant material under these headings. If this is the case, you may have to broaden your search. Since the problem is volcanic dust, you could check general books on dust to see whether cleanup methods for other types of dust could be applied to this problem. For example, there is a lot of information on grain dust. Headings to check could include: "Dust removal," "Dust control," and "Dust explosions." Among the books listed under these headings is *Particulate and Fine Dust Removal: Processes and Equipment,* by Marshall Sitting (1977). To be sure that you have reviewed the most recent research on this topic, you would need to check the recent journal and technical report literature. *Engineering Index* and *Applied Science and Technology* would be good choices for searching the journal literature. There are no subject headings in *Engineering Index* relating specifically to volcanoes. However, you might want to check "Dust control" and "Dust collectors," which are index headings here. A recent journal article listed under "Dust control" is by C. W. Vagelsang, Jr., entitled "Operating Costs of Particulate Removal Schemes," in *Chemical Engineering Progress,* December 1978, pages 55–58. *Applied Science and Technology Index* uses the same subject headings listed in *LC List of Subject Headings.* Although *Engineering Index* does cover some technical reports, you might also want to check the *Monthly Catalog* and *Government Reports Announcements and Index* for additional information. These two government indexes also use the subject headings from the *LC List.* Many technical reports and documents can be found under these headings. An example of a government document is the U.S. Bureau of Mines *Mineral Industry Survey,* "Pumice and Volcanic Cinder" (1978). A sample technical report is "Effect of Water Sprays for Respirable Dust Suppression with a Research Continuous Mining Machine," by Welby G. Courtney, Natesa I. Jayaraman, and Paul Behum.

PATENTS

Patents are very important sources of information for engineers, particularly those engineers who have new design ideas. Patents are often cited in the literature and can contain information not found in books, periodical articles, or technical reports. Thus, knowing what patents are and how to find them is essential for any engineer trying to obtain a patent. This section cannot possibly cover the whole topic of patent searching, so interested readers are referred to books on this subject. It is not necessary that engineers become experts in patent law because larger firms usually engage the services of patent librarians and attorneys to search the patent literature and apply for patents on products. Still, it is important that engineers understand the basic fundamentals of patents and their uses.

Patent Fundamentals

Protection is now offered under five basic categories: *utility patents, design patents, plant patents, trademarks,* and *copyrights.*

UTILITY PATENT A utility patent is a grant giving its owner the right to exclude others from making, using, and selling useful *products, machines, articles of manufacture,* and *compositions of matter.*

DESIGN PATENT A design patent is a grant with protection rights similar to those of utility patents that covers *new, original, ornamental,* and *unobvious designs* for articles of manufacture.

PLANT PATENT A plant patent is a grant with protection rights similar to those of utility patents that covers new varieties of plants that have been *asexually reproduced,* with the exception of tuber-propagated plants and those found in an uncultivated state. These varieties must have been produced by means other than seeds such as by grafting or hybridization.

TRADEMARK A trademark is a legal protection afforded to a wide variety of different *words, names, symbols, configurations, devices,* or any combination of these, which is used to distinguish goods or services from those of others.

COPYRIGHTS A copyright is legal protection afforded to the *creative works* of the artist, either in music, literature, or the aesthetic arts.

Thus, a wide range of things can be legally protected from those who would profit from the creative efforts of others. (One of the most interesting legal decisions of this century was made by the Supreme Court in June 1980. The crux of the decision was that biological forms of life produced in the laboratory are patentable under the category of utility patents. This has stimulated significant research efforts in genetic engineering.)

Not all things can be protected by law. Products of nature, natural laws or principles, and new business methods are examples. Even original inventions are not patentable if they are in public use for a year or more before the patent is applied for.

Table 7.1 lists the protection rights and protection sources for each of the five categories discussed. It is interesting to note that each year, 65 000 utility patents are issued, 50 000 trademark applications are received, and 450 000 copyright registrations are issued.

Patent Searching

Only a selected number of depository libraries receive all patents. Most college and university libraries may not even have a patent collection. However, one can obtain a copy of a patent directly from the patent office or through interlibrary loan service.

There are several ways to learn about new patents. The U.S. Patent and Trademark Office, which grants patents in this country, also provides the

TABLE 7.1   Protection Rights and Protection Sources

	Term of protection	Tests for infringement	Sources of protection
Utility patent	17 years	Making, using, or selling invention claimed in patent	U.S. Patent and Trademark Office
Design patent	$3\frac{1}{2}$, 7, or 14 years, depending on elected term	Making, using or selling design shown in patent drawings	U.S. Patent and Trademark Office
Plant patent	17 years	Reproduction, use, or selling of a plant asexually produced (reproducing by seeds is not patent infringement)	U.S. Patent and Trademark Office or Dept. of Agriculture
Trademark	State terms vary, usually 10 years and renewable; 20 years and renewable for federal registration	Possibility of confusion, mistake, or deception	State registration offices, common law protection, and U.S. Patent and Trademark Office
Copyright	Life of author plus 50 years for works created after Jan. 1, 1978	Copying of subject matter	U.S. Copyright Office

Official Gazette Patents. This weekly publication lists recently issued patents. In it, patents are grouped into three broad categories: general and mechanical, chemical, and electrical. The *Official Gazette* contains a drawing and abstract of each patent.

The *Official Gazette* also publishes an annual index of patentees and subjects. Unfortunately, the subject indexing is not very thorough. Several of the standard periodical indexes include patents and provide much better indexing than that provided in the *Official Gazette,* for example, *Chemical Abstracts* in the area of chemical engineering patents. *Chemical Abstracts* also publishes a special index called a patent concordance. This index correlates patents issued by different countries for the same basic invention. *Physics Abstracts, Electrical and Electronics Abstracts,* and *Science Citation Index* are just a few of the other indexes that also include patents. In addition to these general indexes, which can be found in most college and university libraries, there are a number of specialized patent indexing services that will be found mainly in private industrial libraries.

Patent Structure

Patents may vary greatly in length, but their format is always the same. The number of the patent is given in the upper right-hand corner. This unique patent number is the most important piece of information for identifying and obtaining a patent.

Other important identifying facts are the name of the patentee or inventor, the date the patent was granted, and the date of the original patent application. The body of the patent first summarizes the invention and then describes the invention in detail. This detailed description will often include drawings and design plans. Frequently, in explaining the background for the invention, the patentee will also cite many other related literature and patent references, thus providing a nice review of the literature.

Following the description that makes up the body of a patent is a list of numbered claims that, in very specific language, describes what the inventor is trying to get patented. If one is trying to find out whether an item has already been patented, the claims are the most important section of a patent. The claims of a patent are usually narrower than the information disclosed in a patent. Any information revealed in a patent but not covered by the claims is freely available for public use.

Patents and Records

Engineers must understand the importance of keeping records of their work and, particularly, their ideas. It is not always obvious when something becomes patentable. Other engineers working for competing companies might simultaneously be developing some new or similar process or product that may be patentable and ultimately bring significant return to the company. In these cases prior discovery may have to be proved in the courts in one of the following ways:

- Written, dated, and witnessed disclosure fully describing the invention. It must be clear and understandable to anyone skilled in the particular art.
- Dated and identified documents that describe the steps leading up to the invention, and possibly documents describing the developments following the invention.
- Formal patent application for the invention.
- Witnessed operating model of the invention.

Students will comprehend better the place of patents in industrial R&D management activities by reading the "playlet" of the following section.

THE DRAMA OF PATENTS*

Ignorance of patent practice has rung down the curtain upon many a promising new product. Our playlet tells you the errors to avoid.

*Reprinted by special permission from *Chemical Engineering, Supplementary Readings in Engineering Design*, 1975. By Francis W. Guay, Allied Chemical Corp., Copyright © 1974, by McGraw-Hill, Inc., New York, N.Y. 10020.

The stage is set in the labs of Trylar Plastic Co., a division of giant Stardust Chemicals, Inc.

The Trylar commercial development people are convinced that a new polymer, "polyglot Q," being written up in all the trade periodicals is a natural for them.

Here is the background for our patent play. Although polyglot Q differs quite a bit chemically from Trylar's present commercial polymers, its potential markets are parallel to those for the firm's existing line of high-quality molding compounds. And the technical service department already envisions specialized uses for it.

A preliminary look-see by the R&D people won their enthusiasm. They predicted a relatively easy reaction, and they have had long experience with the principal reactant. Nevertheless, they had the usual reservations as to yields.

Now, let's look at the cast of our drama. Charlie Gogeau, project manager, insists his market-intelligence study, and all the other information he has been culling, indicates a winner in polyglot Q. Most importantly, he has the full support and urging of Dr. Barker, vice-president and technical director. Barker is an avowed management-by-objectives leader—the kind of task-master who loudly claims, "Who says we can't be on stream in five months?"

Needless to say, the new president of Trylar Plastics Co. would be thrilled with the glory of announcing this brand-new product to the trade. In fact, from the flowing predictions he's heard for polyglot Q, he's already thinking his men had better come through with perfecting it or there's a distinct possibility that some heads might roll. "After all," muses the president, "how long has it been since Trylar has had a really new product—except by acquisition. What a coup that would be!"

Well, the stage is finally set, but Act I is already ending.

Act II, one month later, finds Charlie Gogeau's commercial development team and the R&D exploratory and feasibility team at a no-holds-barred meeting. Quite unexpectedly, Dr. Barker walks in.

DR. BARKER: Charlie, I've been getting your weekly reports, and I see all you people are making real progress in staffing up. I'm also glad to see that your enthusiasm is still high—even after your refined market and cost analyses. Now what about the preliminary R&D lab tests, how do they look?

CHARLIE: Well, Dr. Barker, because of the sparse information in scientific reports and the trade periodicals, the R&D people are encountering difficulties in making polyglot Q.

DR. BARKER: What about patents? Can't we learn something from them? And by the way, we know duCro is working on a process. Can we get a sample?

CHARLIE: Well, Fred Gray from the patent department says he has uncovered a batch of about fifty U.S. patents dealing with polyglot-type polymers, but he wants to get together with me to talk about them. Says he has to know more about the specific process that we contemplate.

DR. BARKER: Aw, hell! Those patent guys always want specifics. Tell you what, find out if there are any product patents first. I'm not worried about process patents. Does duCro have anything?

CHARLIE: Yes, duCro has quite a few patents, but I don't really know what they are yet.

DR. BARKER: (*pontificating*) Product patents can be trouble, but I haven't seen a process patent yet that you can't get around. You guys keep up the good work. And don't forget, we've only got until September first to make our final decision. Remember, the big boss and I are expecting great things from you. After all, it's been my policy to hire only sharp, wide-eyed, and aggressive research and development talent over the past few years. (*Exit Dr. Barker*)

Charlie and his patent attorney, Fred Gray, now have their tête-à-tête, and the upshot is that they find that thirty-eight of the fifty patents are only remotely pertinent and the remaining twelve, while they deal with polyglot-type polymer chemistry, do not deal directly with polyglot Q, nor do they specifically cover the particular reactants that R&D is trying out.

CHARLIE: How come there isn't anything on polyglot Q, Fred? The reaction and the product have both been disclosed for some time. Why it seems that there hasn't been a month gone by in the past year without some scientific publication commenting on this new polymer. Furthermore, from what is generally known in the polymer chemistry field, it wouldn't take a genius to bring the specific reactants together and make the reaction go.

FRED GRAY: Well, my first impression after reading this bundle of patents is that it's quite possible (assuming that we've got all the patents here) to conceive of duCro or someone else working on getting some pretty significant patent coverage on both product and process.

Tell you what I'll do. I'll have a search conducted at The Hague.* It's quite possible that certain European countries have issued patents bearing on polyglot Q. Countries like France and Belgium, for example, have no examining procedure, and patents are published considerably earlier there than their counterparts in the U.S.

CHARLIE: Fine. But hurry it up as much as possible. Remember, we've a deadline to beat. (CURTAIN)

---

*Institute International des Brevets, The Hague, Netherlands.

The time is several weeks later in Charlie Gogeau's office. The phone rings. It's Fred Gray on the other end of the line.

FRED: Charlie, listen to this. My German isn't very good, but The Hague found a Dutch patent publication to duCro that seems to read right smack on polyglot Q. Besides that, it has fifty-three examples, and the patent attorney who wrote it went overboard with product equivalents and process variables.

CHARLIE: When can I see it?

FRED: Look, I'll shoot it off to you; once you've read it, buzz me and we'll get together. Meanwhile, I'll order the U.S. priority document.

CHARLIE: The what?

FRED: I'll explain it to you when we sit down together. Believe me, it'll be fun from here on in.

At this juncture, the patent attorney knows full well that his work is really cut out for him. He knows that the client is sure to raise a host of theoretical arguments against the patentability, the scope, the merit, the everything of duCro's patent application on the alleged novel process for making polyglot Q.

The patent attorney, however, is faced with the job of guessing just what kind of patent duCro is likely to get. On the basis of his knowledge of the operation of the Patent Office and his background knowledge of previous work in the polymer field, he must evaluate just what claims the U.S. Patent Office is likely to allow. Then, on the basis of this hypothetical patent, he has to decide if Trylar Plastics' product and process will infringe.

Furthermore, he has to remember that the whole plastics industry is interested in the very desirable properties of this new plastic. A number of companies have spent money evaluating it. Trylar Plastics, however, is committed to its manufacture.

Complications Set In

In time, Fred Gray received a copy of the U.S. priority document that duCro had filed in the Dutch patent office. Why was it there? This was what happened:

duCro had first prepared a patent application that was filed in the U.S. Patent Office. duCro later filed a corresponding application in the Dutch patent office, accompanied by a certified copy of the earlier-filed U.S. application (the so-called "priority document"). This established an early "date-of-filing" that would apply even though the formal Dutch application was filed later. After the Dutch application was published, the priority document became available to the public.

This means that although the U.S. Patent Office holds applications in secrecy, it is often possible to get a copy of the application by such a device.

Well, Fred Gray's German might have been rusty, but when he read the priority document, which was in plain English, it made his head spin. All he saw were problems.

Charlie Gogeau, on the other hand, together with his associates, was delirious. Just think, there were examples galore, preferred procedures (including reactants and detailed reaction conditions), comparative data, caveats, etc., etc.

Looking at the same papers, Fred Gray saw generic product claims as broad as the Atlantic Ocean, subgeneric claims as broad as the Mediterranean, even subsubgeneric product claims, and, of course, species claims—all generously documented in a specification containing a string of analogs and equivalents for each reactant.

Within a fortnight, the R&D exploratory and feasibility team had zeroed in on polyglot Q, and the polymer was everything its press notices had said it was. It was a beautiful polymer. The Commercial Development people were all fired up; Charlie Gogeau was ecstatic. Needless to say, this enthusiasm rubbed off on Dr. Barker, vice-president and technical director—and from him to the president of Trylar Plastics Co.

The September first date was nearing—in fact only seventeen days off. Dr. Barker called a progress review meeting to pull all the findings and recommendations together.

The meeting went along swimmingly—that is, until Dr. Barker asked Fred to report on the patent situation.

As tactfully as he knew how, Fred told the assembled scientists that it was his legal opinion after studying both duCro's patent application and the prior art, that duCro would be granted fairly broad product and process claims. Also, the process that R&D had just described would infringe these claims.

In view of this, it was his suggestion that processes differing materially from duCro's be explored. Also, any products of the polyglot polymer type that might be made would have to differ essentially from polyglot Q and its analogs as discussed in the duCro application. As such products or processes were developed, each would have to be carefully scrutinized to see if there were any infringement or potential infringement problems.

Dr. Barker Gets Hot Under the Collar

Dr. Barker blew his cool, and the reaction spread around the room. Why was it, Dr. Barker asked, that every time they ran into the other fellow's patent they were told they couldn't go ahead because it was a valid patent, while, on the other hand, their own patents weren't held in such high esteem?

Fred quickly pointed out that this was an unfair accusation. In the past two months, he had cleared several proposed processes that clearly infringed a competitor's patent, because he had concluded (after studying the prior art) that the claims were too broad and hence invalid.

In addition, Fred reminded Dr. Barker that while it was the responsibility of counsel to offer a legal opinion, whether to go ahead with the process or not was strictly the client's decision.

Charlie Gogeau went on record with the opinion that condensation reactions of the general type claimed in duCro's application were as old as the hills, and to apply them to the known reactants in the application would be obvious to the skilled chemist.

Fred countered by asking him why, according to the application, conventional condensation catalysts didn't work.

Needless to say, the meeting lasted through lunch, and then some, proving a very tiring experience to Fred. He did manage to explain to the group, when the dust settled a bit, that it would be really worthwhile for them to put their heads together to come up with alternative reactants and catalysts for preparing analogous products having, hopefully, materially different properties.

Fred suggested to the assembled researchers that if the reaction was obvious, as suggested in hindsight, why not put a bit of foresight to work in the interest of getting away from the patent. After all, he pointed out, one of the healthiest consequences of a patent is that it stimulates other ideas and inventions.

Fred also suggested that it would be worthwhile, in view of the intense interest of Trylar Plastics, to make a definitive search of the scientific literature. Since everyone claimed that the reaction system wasn't unique, it might have been suggested or described in some obscure periodical, or even in a doctoral thesis, or the like. Even though he didn't appreciate the potential properties of polyglot Q, someone, somewhere, might have implicitly or explicitly suggested the reaction, or the products, or the catalyst. Such publication might invalidate duCro's patent.

Dr. Barker ordered the search immediately.

A week passed and some fifteen suggestions for possible alternative routes to a polyglot-Q-type polymer came across Fred Gray's desk. At first glance, some seemed quite exotic, many looked dangerously close to the types in duCro's application—as for the rest, well. . . .

A week later, and before he had had a chance to really study and draw conclusions from the first fifteen, six more came in carrying a "please hurry" note, referring to all twenty-one suggestions.

At least five more meetings ensued in the next two weeks, in which Fred Gray was questioned, cross-questioned, requestioned and re-cross-questioned for the position he took on the merits of duCro's application.

Fred had turned thumbs down on seventeen of the twenty-one alternative routes proposed, for the reason that they either involved art-recognized equivalents or reactants implicitly suggested in the duCro application, or for other similar reasons. Fred's reasons in each case were subject to the same rigorous questioning.

In addition, he had to present his position at meetings right up through the chain of command. This series of meetings finally culminated in a sort of star chamber session at headquarters.

This was a command performance in every sense of the word, with virtually every person connected with the polyglot Q project in attendance: a white-haired executive vice-president of Stardust Chemicals, Inc., on whose shoulders rested the ultimate decision; the top echelon of Trylar Plastics—including, of course, the good Dr. Barker; the chief corporate counsel; R&D and Commercial Development along with, naturally, Charlie Gogeau; and there in the flesh sat everybody's "friend," good old Fred Gray. He was easily distinguishable as a villain because of the black "Hathaway" patch over his right eye (the result of a weekend mishap when a particle of metal got into his eye).

R&D very succinctly presented its encouraging findings on polyglot Q as well as its minor successes and frustrations with regard to the backup alternative routes. Commercial Development's prognosis for polyglot Q had not changed; it was still glowing. Like R&D, its presentation was well documented and easily comprehended; like R&D, it could not sing the praises of the backup routes.

Charlie Gogeau did himself proud with his "recap" of the total polyglot Q project, fully emphasizing what this polymer with its outstanding properties could do for Trylar Plastics Co. and, of course, Stardust Chemicals, Inc. Dr. Barker chimed in with a few well-chosen phrases to sort of crystallize the perspective Charlie Gogeau was attempting to establish.

Except for a few requests for clarification, the white-haired corporate executive vice-president sat very attentively, seemingly recollecting similar experiences earlier in his career.

And now, last but not least, we present . . . the patent situation. Introducing Fred Gray. All eyes were on old one-eye. If ever there was a villain!

A Fatal Prognosis

Throughout the meeting Fred was mentally toying with several ways of telling the patient the diagnosis. Should he work up to it by discussing the various symptoms, diagnostic tools, precautionary steps taken to avoid mistakes, and finally, the chances of recovery? Or should he point blank tell him that he has cancer and the prognosis is bleak? Either approach he felt would generate a wave of questioning, so why postpone the reaction? Besides that, Fred's bad eye was killing him at this point.

In as concise and unambiguous a manner as he could, he laid out the issue and promptly presented his conclusion:

"duCro is expected to procure a strong patent position on polyglot with both broad product claims and, of course, process claims. The process of choice described by our R&D is also the preferred one in the duCro application and, on the basis of the prior art we have so far uncovered, should have a very good chance of standing up in court in the event of litigation. Any court would most definitely take cognizance in its decision of the breakthrough represented by duCro's new polymer, the commercial interest it has stimulated, and the notoriety it has received; in the language of these decisions, the court looks for the patentee's 'contribution.'"

As to the possibility of litigation, Fred merely asked if there was anyone in the room who wouldn't sue if he owned this invention. No one spoke up.

With that, he took a neatly folded white handkerchief from his pocket, lifted the black eye patch, and gently padded the eye. He had spoken his piece and now awaited the grilling.

A few innocuous questions were asked, one of them from the white-haired corporate executive vice-president who wanted to know if duCro had been approached for a license. It had been; it had refused. Then suddenly the meeting lagged for a few seconds. What happened? thought Fred. I came here fully armed for combat but. . . .

Just then the soft-spoken, white-haired old gentleman stood up and said, "Gentlemen, are there any more reports?" He was assured the whole story had been told. "I was very impressed with the content and clarity of your reports, and I can certainly appreciate the frustrations a project of this magnitude entails and the hours of work that go into it. In my opinion, however, we have apparently not been able to shake duCro's position and I agree with counsel that we are inviting a lawsuit. While we were successful in a previous court case in this field, you must remember that there we had developed our own unique process and we were able to throw real big rocks at the patent. We honestly felt it was invalid and had substantial art to present.

"For those of you who haven't been in a real patent lawsuit, let me tell you it is bloody even when you think your case is a very good one—not to mention its cost. Gentlemen, if you'll excuse me, I've got another meeting that I am already late for."

Without further adieu, everyone quietly drifted out of the room in what seemed to be in every direction. Fred had a luncheon engagement that started out with a double martini, straight up, with two olives!

What Does It Mean?

Success story for counsel? Absolutely not. Everyone at that meeting was a bit wiser. The object lesson is obvious:

Good business and research planning must include a thorough and timely study of the patent monopolies granted inventors for divulging their discoveries.

Any management that makes light of or ignores the vast network of unexpired or soon-to-be issued patents probably has not previously encountered a frustration of the type described above, and is probably not properly protecting its own discoveries. Obviously, a very fertile area in which to try to establish property rights by way of patents could be found in the alternative routes mentioned above that R&D had researched.

While copying with a view to improving is a time-honored research objective, it is absolutely necessary for management to appreciate that the originators of the basic invention either have sought or are diligently seeking patent coverage commensurate in scope with their discovery.

The philosophy that most patents are not worth the paper they are written on ignores the fact that individuals and corporations have amassed fortunes by virtue of having patented, nationally and internationally, their better mousetraps. On the other hand, no one can deny that a number of patents are acquired strictly for their nuisance value and that they only meagerly promote the progress of science.

Whether or not you consider reprehensible this practice of harassing your competitors or establishing a body of so-called "defensive" patents, it would be foolhardy to leap headlong into a market without first making a judgment as to the merit of all relevant patents and patent applications.

In other words, if it's a product you want to make, use, or sell, you must first make sure that the product or the process you contemplate for making it is not the property of another, or is not about to be.

In any project, a thorough investigation for potential patent impediments should be scheduled early and given equal weight with other studies, such as technical feasibility, market research, cost analyses, and so on. Should any one of these go sour, the project becomes a failure.

In the last analysis, our patent system is intended to reward the true inventor with a patent containing claims that are coextensive in scope with the breadth of the invention, that is, the contribution to the art. Counsel likes nothing more than to procure a valuable patent for a valuable invention. Judging from the disclosure in the polyglot Q patent application, the assignee, duCro, recognized the full potential its inventor's discovery had, and counsel responded with a well-drafted and well-documented application. Fred Gray, likewise, appreciated the fact that this was not boilerplate; it was a genuine effort to protect what appeared to be a significant scientific contribution by procuring patent coverage that would survive the close scrutiny of competitors and, of course, the courts, in the event of litigation.

If this incident in the whole scheme of scientific innovation and development bolsters anyone's confidence in a patent system that is under continuing heavy pressure from genuine and misguided reformers alike, it will have been worthwhile. While it is still possible in the United States to acquire a meritorious patent grant for a meritorious invention, let us be alert to those who—in the interest of streamlining the patent system—might just inadvertently destroy it. In short, it is very possible that their preoccupation with shortcomings will precipitate changes that are totally inconsistent with the original purpose of the patent system as envisioned by our founding fathers— "to promote the progress of science . . . by securing for limited times to . . . inventors the exclusive right to their . . . discoveries."

PROBLEMS

7.1 Summarize the contents of U.S. patent number 3,780,345.

7.2 A person named Chirico obtained a patent on heat exchangers sometime in the late 1970s. Find out the exact patent number so that the patent can be ordered.

7.3 Find three information sources that describe the designs of space simulators for testing unmanned spacecraft.

7.4 Exxon is currently involved in government-sponsored research on coal liquefaction. Explain how you would find out about any recent progress in their research.

7.5 Find the addresses of three companies that make neon signs.

7.6 Find the different definitions for the word "block" as used in data processing, design engineering, mining engineering, and petroleum engineering.

7.7 What is the Haber-Bosch process and what product is synthesized by means of this process?

7.8 List four chemical synonyms for aspirin. How is aspirin usually prepared?

7.9 What is the difference between the ANSI and the IEEE Standards for Performance Specification for Reactor Emergency Radiological Monitoring Instrumentation?

7.10 Find citations to articles in engineering journals about recent salaries for engineers. Which index did you use to find these citations?

7.11 Find citations to journal articles on petroleum refinery equipment. Find citations to journal articles on artificial satellites used for communication applications.

Appendix A
Conversion Factors
and Other Constants

Conversion Factors

The following tables express the definitions of miscellaneous units of measure as exact numerical multiples of coherent SI units. They also provide multiplying factors for converting numbers and miscellaneous units to corresponding new numbers and SI units.

An asterisk follows each number that expresses an exact definition. For example, the entry "2.54* $\times$ 10^{-2}" expresses the fact that 1 inch equals 2.54 $\times$ 10^{-2} meter exactly by definition. Numbers not followed by an asterisk are only approximate representations of definitions or are the results of physical measurements.

The conversion factors are listed alphabetically and by physical quantity. The listing by physical quantity includes only relationships that are frequently encountered. It deliberately omits the great multiplicity of combinations of units that are used for more specialized purposes. Conversion factors for combinations of units are easily generated from numbers given in the alphabetical listing by the technique of direct substitution or by the rules for manipulating units.

Alphabetical Listing

To convert from	To	Multiply by
abampere	ampere	1.00* $\times$ 10^1
abcoulomb	coulomb	1.00* $\times$ 10^1
abfarad	farad	1.00* $\times$ 10^9
abhenry	henry	1.00* $\times$ 10^{-9}
abmho	siemens	1.00* $\times$ 10^9
abohm	ohm	1.00* $\times$ 10^{-9}
abvolt	volt	1.00* $\times$ 10^{-8}

Reprinted from NASA publication SP 7012, *The International System of Units, Physical Constants and Conversion Factors*, by E. A. Mechtly, 2nd revision, 1973.

Alphabetical Listing (*continued*)

To convert from	To	Multiply by
acre	meter^2	4.046 856 422 4* $\times$ 10^3
angstrom	meter	1.00* $\times$ 10^{-10}
are	meter^2	1.00* $\times$ 10^2
astronomical unit (IAU)	meter	1.496 00 $\times$ 10^{11}
astronomical unit (radio)	meter	1.495 978 9 $\times$ 10^{11}
atmosphere	newton/meter^2	1.013 25* $\times$ 10^5
bar	newton/meter^2	1.00* $\times$ 10^5
barn	meter^2	1.00* $\times$ 10^{-28}
barrel (petroleum, 42 gallons)	meter^3	1.589 873 $\times$ 10^{-1}
barye	newton/meter^2	1.00* $\times$ 10^{-1}
board foot (1′ $\times$ 1′ $\times$ 1″)	meter^3	2.359 737 216* $\times$ 10^{-3}
British thermal unit:		
(IST before 1956)	joule	1.055 04 $\times$ 10^3
(IST after 1956)	joule	1.055 056 $\times$ 10^3
British thermal unit (mean)	joule	1.055 87 $\times$ 10^3
British thermal unit (thermochemical)	joule	1.054 350 $\times$ 10^3
British thermal unit (39°F)	joule	1.059 67 $\times$ 10^3
British thermal unit (60°F)	joule	1.054 68 $\times$ 10^3
bushel (U.S.)	meter^3	3.523 907 016 688* $\times$ 10^{-2}
cable	meter	2.194 56* $\times$ 10^2
caliber	meter	2.54* $\times$ 10^{-4}
calorie (International Steam Table)	joule	4.1868
calorie (mean)	joule	4.190 02
calorie (thermochemical)	joule	4.184*
calorie (15°C)	joule	4.185 80
calorie (20°C)	joule	4.181 90
calorie (kilogram, International Steam Table)	joule	4.1868 $\times$ 10^3
calorie (kilogram, mean)	joule	4.190 02 $\times$ 10^3
calorie (kilogram, thermochemical)	joule	4.184* $\times$ 10^3
carat (metric)	kilogram	2.00* $\times$ 10^{-4}
Celsius (temperature)	kelvin	$t_k = t_c + 273.15$
centimeter of mercury (0°C)	newton/meter^2	1.333 22 $\times$ 10^3
centimeter of water (4°C)	newton/meter^2	9.806 38 $\times$ 10^1
chain (engineer or ramden)	meter	3.048* $\times$ 10^1
chain (surveyor or gunter)	meter	2.011 68* $\times$ 10^1
circular mil	meter^2	5.067 074 8 $\times$ 10^{-10}
cord	meter^3	3.624 556 3
cubit	meter	4.572* $\times$ 10^{-1}
cup	meter^3	2.365 882 365* $\times$ 10^{-4}
curie	disintegration/second	3.70* $\times$ 10^{10}
day (mean solar)	second (mean solar)	8.64* $\times$ 10^4
day (sidereal)	second (mean solar)	8.616 409 0 $\times$ 10^4
degree (angle)	radian	1.745 329 251 994 3 $\times$ 10^{-2}
denier (international)	kilogram/meter	1.00* $\times$ 10^{-7}
dram (avoirdupois)	kilogram	1.771 845 195 312 5* $\times$ 10^{-3}

Alphabetical Listing (*continued*)

To convert from	To	Multiply by
dram (troy or apothecary)	kilogram	$3.887\ 934\ 6^* \times 10^{-3}$
dram (U.S. fluid)	meter^3	$3.696\ 691\ 195\ 312\ 5^* \times 10^{-6}$
dyne	newton	$1.00^* \times 10^{-5}$
electron volt	joule	$1.602\ 191\ 7 \times 10^{-19}$
erg	joule	$1.00^* \times 10^{-7}$
Fahrenheit (temperature)	kelvin	$t_K = (5/9)\ (t_F + 459.67)$
Fahrenheit (temperature)	Celsius	$t_C = (5/9)\ (t_F - 32)$
faraday (based on carbon-12)	coulomb	$9.648\ 70 \times 10^4$
faraday (chemical)	coulomb	$9.649\ 57 \times 10^4$
faraday (physical)	coulomb	$9.652\ 19 \times 10^4$
fathom	meter	$1.828\ 8^*$
fermi (femtometer)	meter	$1.00^* \times 10^{-15}$
fluid ounce (U.S.)	meter^3	$2.957\ 352\ 956\ 25^* \times 10^{-5}$
foot	meter	$3.048^* \times 10^{-1}$
foot (U.S. survey)	meter	$1200/3937^*$
foot (U.S. survey)	meter	$3.048\ 006\ 096 \times 10^{-1}$
foot of water (39.2°F)	newton/meter^2	$2.988\ 98 \times 10^3$
footcandle	lumen/meter^2	$1.076\ 391\ 0 \times 10^1$
footlambcrt	candela/meter^2	$3.426\ 259$
free fall, standard	meter/second^2	$9.806\ 65^*$
furlong	meter	$2.011\ 68^* \times 10^2$
gal (galileo)	meter/second^2	$1.00^* \times 10^{-2}$
gallon (U.K. liquid)	meter^3	$4.546\ 087 \times 10^{-3}$
gallon (U.S. dry)	meter^3	$4.404\ 883\ 770\ 86^* \times 10^{-3}$
gallon (U.S. liquid)	meter^3	$3.785\ 411\ 784^* \times 10^{-3}$
gamma	tesla	$1.00^* \times 10^{-9}$
gauss	tesla	$1.00^* \times 10^{-4}$
gilbert	ampere turn	$7.957\ 747\ 2 \times 10^{-1}$
gill (U.K.)	meter^3	$1.420\ 652 \times 10^{-4}$
gill (U.S.)	meter^3	$1.182\ 941\ 2 \times 10^{-4}$
grad	degree (angular)	$9.00^* \times 10^{-1}$
grad	radian	$1.570\ 796\ 3 \times 10^{-2}$
grain	kilogram	$6.479\ 891^* \times 10^{-5}$
gram	kilogram	$1.00^* \times 10^{-3}$
hand	meter	$1.016^* \times 10^{-1}$
hectare	meter^2	$1.00^* \times 10^4$
hogshead (U.S.)	meter^3	$2.384\ 809\ 423\ 92^* \times 10^{-1}$
horsepower (550 ft·lb$_f$/s)	watt	$7.456\ 998\ 7 \times 10^2$
horsepower (boiler)	watt	$9.809\ 50 \times 10^3$
horsepower (electric)	watt	$7.46^* \times 10^2$
horsepower (metric)	watt	$7.354\ 99 \times 10^2$
horsepower (U.K.)	watt	7.457×10^2
horsepower (water)	watt	$7.460\ 43 \times 10^2$
hour (mean solar)	second (mean solar)	$3.60^* \times 10^3$
hour (sidereal)	second (mean solar)	$3.590\ 170\ 4 \times 10^3$
hundredweight (long)	kilogram	$5.080\ 234\ 544^* \times 10^1$

Alphabetical Listing (*continued*)

To convert from	To	Multiply by
hundredweight (short)	kilogram	$4.535\ 923\ 7^* \times 10^1$
inch	meter	$2.54^* \times 10^{-2}$
inch of mercury (32°F)	newton/meter^2	$3.386\ 389 \times 10^3$
inch of mercury (60°F)	newton/meter^2	$3.376\ 85 \times 10^3$
inch of water (39.2°F)	newton/meter^2	$2.490\ 82 \times 10^2$
inch of water (60°F)	newton/meter^2	2.4884×10^2
kilocalorie (International Steam Table)	joule	$4.186\ 8 \times 10^3$
kilocalorie (mean)	joule	$4.190\ 02 \times 10^3$
kilocalorie (thermochemical)	joule	$4.184^* \times 10^3$
kilogram force (kg$_f$)	newton	$9.806\ 65^*$
kilogram mass	kilogram	1.00^*
kilopound force	newton	$9.806\ 65^*$
kip	newton	$4.448\ 221\ 615\ 260\ 5^* \times 10^3$
knot (international)	meter/second	$5.144\ 444\ 444 \times 10^{-1}$
lambert	candela/meter^2	$1/\pi^* \times 10^4$
lambert	candela/meter^2	$3.183\ 098\ 8 \times 10^3$
langley	joule/meter^2	$4.184^* \times 10^4$
lb$_f$ (pound force, avoirdupois)	newton	$4.448\ 221\ 615\ 260\ 5^*$
lb$_m$ (pound mass, avoirdupois)	kilogram	$4.535\ 923\ 7^* \times 10^{-1}$
league (U.K. nautical)	meter	$5.559\ 552^* \times 10^3$
league (international nautical)	meter	$5.556^* \times 10^3$
league (statute)	meter	$4.828\ 032^* \times 10^3$
light year	meter	$9.460\ 55 \times 10^{15}$
link (engineer or ramden)	meter	$3.048^* \times 10^{-1}$
link (surveyor or gunter)	meter	$2.011\ 68^* \times 10^{-1}$
liter	meter^3	$1.00^* \times 10^{-3}$
lux	lumen/meter^2	1.00^*
maxwell	weber	$1.00^* \times 10^{-8}$
meter	wavelengths Kr-86	$1.650\ 763\ 73^* \times 10^6$
micron	meter	$1.00^* \times 10^{-6}$
mil	meter	$2.54^* \times 10^{-5}$
mile (international nautical)	meter	$1.852^* \times 10^3$
mile (U.K. nautical)	meter	$1.853\ 184^* \times 10^3$
mile (U.S. nautical)	meter	$1.852^* \times 10^3$
mile (U.S. statute)	meter	$1.609\ 344^* \times 10^3$
millibar	newton/meter^2	$1.00^* \times 10^2$
millimeter of mercury (0°C)	newton/meter^2	$1.333\ 224 \times 10^2$
minute (angle)	radian	$2.908\ 882\ 086\ 66 \times 10^{-4}$
minute (mean solar)	second (mean solar)	$6.00^* \times 10^1$
minute (sidereal)	second (mean solar)	$5.983\ 617\ 4 \times 10^1$
month (mean calendar)	second (mean solar)	$2.628^* \times 10^6$
nautical mile (international)	meter	$1.852^* \times 10^3$
nautical mile (U.K.)	meter	$1.853\ 184^* \times 10^3$
nautical mile (U.S.)	meter	$1.852^* \times 10^3$
oersted	ampere/meter	$7.957\ 747\ 2 \times 10^1$

Alphabetical Listing (*continued*)

To convert from	To	Multiply by
ounce force (avoirdupois)	newton	$2.780\ 138\ 5 \times 10^{-1}$
ounce mass (avoirdupois)	kilogram	$2.834\ 952\ 312\ 5^* \times 10^{-2}$
ounce mass (troy or apothecary)	kilogram	$3.110\ 347\ 68^* \times 10^{-2}$
ounce (U.S. fluid)	meter^3	$2.957\ 352\ 956\ 25^* \times 10^{-5}$
pace	meter	$7.62^* \times 10^{-1}$
parsec (IAU)	meter	$3.085\ 7 \times 10^{16}$
pascal	newton/meter^2	1.00^*
peck (U.S.)	meter^3	$8.809\ 767\ 541\ 72^* \times 10^{-3}$
pennyweight	kilogram	$1.555\ 173\ 84^* \times 10^{-3}$
perch	meter	5.0292^*
phot	lumen/meter^2	1.00×10^4
pica (printer's)	meter	$4.217\ 517\ 6^* \times 10^{-3}$
pint (U.S. dry)	meter^3	$5.506\ 104\ 713\ 575^* \times 10^{-4}$
pint (U.S. liquid)	meter^3	$4.731\ 764\ 73^* \times 10^{-4}$
point (printer's)	meter	$3.514\ 598^* \times 10^{-4}$
poise	newton second/meter^2	$1.00^* \times 10^{-1}$
pole	meter	5.0292^*
pound force (lb$_f$, avoirdupois)	newton	$4.448\ 221\ 615\ 260\ 5^*$
pound mass (lb$_m$, avoirdupois)	kilogram	$4.535\ 923\ 7^* \times 10^{-1}$
pound mass (troy or apothecary)	kilogram	$3.732\ 417\ 216^* \times 10^{-1}$
poundal	newton	$1.382\ 549\ 543\ 76^* \times 10^{-1}$
quart (U.S. dry)	meter^3	$1.101\ 220\ 942\ 715^* \times 10^{-3}$
quart (U.S. liquid)	meter^3	$9.463\ 592\ 5 \times 10^{-4}$
rad (radiation dose absorbed)	joule/kilogram	$1.00^* \times 10^{-2}$
Rankine (temperature)	kelvin	$t_K = (5/9)t_R$
rayleigh (rate of photon emission)	1/second meter^2	$1.00^* \times 10^{10}$
rhe	meter^2/newton second	$1.00^* \times 10^1$
rod	meter	5.0292^*
roentgen	coulomb/kilogram	$2.579\ 76^* \times 10^{-4}$
rutherford	disintegration/second	$1.00^* \times 10^6$
scruple (apothecary)	kilogram	$1.295\ 978\ 2^* \times 10^{-3}$
second (angle)	radian	$4.848\ 136\ 811 \times 10^{-6}$
second (ephemeris)	second	$1.000\ 000\ 000$
second (mean solar)	second (ephemeris)	Consult American Ephemeris and Nautical Almanac
second (sidereal)	second (mean solar)	$9.972\ 695\ 7 \times 10^{-1}$
section	meter^2	$2.589\ 988\ 110\ 336^* \times 10^6$
shake	second	1.00×10^{-8}
skein	meter	$1.097\ 28^* \times 10^2$
slug	kilogram	$1.459\ 390\ 29 \times 10^1$
span	meter	$2.286^* \times 10^{-1}$
statampere	ampere	$3.335\ 640 \times 10^{-10}$
statcoulomb	coulomb	$3.335\ 640 \times 10^{-10}$
statfarad	farad	$1.112\ 650 \times 10^{-12}$
stathenry	henry	$8.987\ 554 \times 10^{11}$
statohm	ohm	$8.987\ 554 \times 10^{11}$

Alphabetical Listing (*continued*)

To convert from	To	Multiply by
statute mile (U.S.)	meter	$1.609\ 344^* \times 10^3$
statvolt	volt	$2.997\ 925 \times 10^2$
stere	meter^3	1.00^*
stilb	candela/meter^2	1.00×10^4
stoke	meter^2/second	$1.00^* \times 10^{-4}$
tablespoon	meter^3	$1.478\ 676\ 478\ 125^* \times 10^{-5}$
teaspoon	meter^3	$4.928\ 921\ 593\ 75^* \times 10^{-6}$
ton (assay)	kilogram	$2.916\ 666\ 6 \times 10^{-2}$
ton (long)	kilogram	$1.016\ 046\ 908\ 8^* \times 10^3$
ton (metric)	kilogram	$1.00^* \times 10^3$
ton (nuclear equivalent of TNT)	joule	4.20×10^9
ton (register)	meter^3	$2.831\ 684\ 659\ 2^*$
ton (short, 2 000 pound)	kilogram	$9.071\ 847\ 4^* \times 10^2$
tonne	kilogram	$1.00^* \times 10^3$
torr (0°C)	newton/meter^2	$1.333\ 22 \times 10^2$
township	meter^2	$9.323\ 957\ 2 \times 10^7$
unit pole	weber	$1.256\ 637 \times 10^{-7}$
yard	meter	$9.144^* \times 10^{-1}$
year (calendar)	second (mean solar)	$3.1536^* \times 10^7$
year (sidereal)	second (mean solar)	$3.155\ 815\ 0 \times 10^7$
year (tropical)	second (mean solar)	$3.155\ 692\ 6 \times 10^7$
year 1900, tropical, Jan., day 0, hour 12	second (ephemeris)	$3.155\ 692\ 597\ 47^* \times 10^7$

Listing by Physical Quantity

To convert from	To	Multiply by
Acceleration		
foot/second^2	meter/second^2	$3.048^* \times 10^{-1}$
free fall, standard	meter/second^2	$9.806\ 65^*$
gal (galileo)	meter/second^2	$1.00^* \times 10^{-2}$
inch/second^2	meter/second^2	$2.54^* \times 10^{-2}$
Area		
acre	meter^2	$4.046\ 856\ 422\ 4^* \times 10^3$
are	meter^2	$1.00^* \times 10^2$
barn	meter^2	$1.00^* \times 10^{-28}$
circular mil	meter^2	$5.067\ 074\ 8 \times 10^{-10}$
foot^2	meter^2	$9.290\ 304^* \times 10^{-2}$
hectare	meter^2	$1.00^* \times 10^4$
inch^2	meter^2	$6.4516^* \times 10^{-4}$
mile^2 (U.S. statute)	meter^2	$2.589\ 988\ 110\ 336^* \times 10^6$

Reprinted from NASA publication SP 7012, *The International System of Units, Physical Constants and Conversion Factors*, by E. A. Mechtly, 2nd revision, 1973.

Listing by Physical Quantity (*continued*)

To convert from	To	Multiply by
Area		
section	meter^2	2.589 988 110 336* $\times$ 10^6
township	meter^2	9.323 957 2 $\times$ 10^7
yard^2	meter^2	8.361 273 6* $\times$ 10^{-1}
Density		
gram/centimeter^3	kilogram/meter^3	1.00* $\times$ 10^3
lb$_m$/in.^{3}	kilogram/meter^3	2.767 990 5 $\times$ 10^4
lb$_m$/ft^3	kilogram/meter^3	1.601 846 3 $\times$ 10^1
slug/foot^3	kilogram/meter^3	5.153 79 $\times$ 10^2
Energy		
British thermal unit:		
(IST before 1956)	joule	1.055 04 $\times$ 10^3
(IST after 1956)	joule	1.055 056 $\times$ 10^3
British thermal unit (mean)	joule	1.055 87 $\times$ 10^3
British thermal unit (thermochemical)	joule	1.054 350 $\times$ 10^3
British thermal unit (39°F)	joule	1.059 67 $\times$ 10^3
British thermal unit (60°F)	joule	1.054 68 $\times$ 10^3
calorie (International Steam Table)	joule	4.1868
calorie (mean)	joule	4.190 02
calorie (thermochemical)	joule	4.184*
calorie (15°C)	joule	4.185 80
calorie (20°C)	joule	4.181 90
calorie (kilogram, International Steam Table)	joule	4.1868 $\times$ 10^3
calorie (kilogram, mean)	joule	4.190 02 $\times$ 10^3
calorie (kilogram, thermochemical)	joule	4.184* $\times$ 10^3
electron volt	joule	1.602 191 7 $\times$ 10^{-19}
erg	joule	1.00* $\times$ 10^{-7}
foot lb$_f$	joule	1.355 817 9
foot poundal	joule	4.214 011 0 $\times$ 10^{-2}
joule (international of 1948)	joule	1.000 165
kilocalorie (International Steam Table)	joule	4.1868 $\times$ 10^3
kilocalorie (mean)	joule	4.190 02 $\times$ 10^3
kilocalorie (thermochemical)	joule	4.184* $\times$ 10^3
kilowatt-hour	joule	3.60* $\times$ 10^6
kilowatt-hour (international of 1948)	joule	3.600 59 $\times$ 10^6
ton (nuclear equivalent of TNT)	joule	4.20 $\times$ 10^9
watt-hour	joule	3.60* $\times$ 10^3
Energy/area time		
Btu (thermochemical)/foot^2 second	watt/meter^2	1.134 893 1 $\times$ 10^4
Btu (thermochemical)/foot^2 minute	watt/meter^2	1.891 488 5 $\times$ 10^2
Btu (thermochemical)/foot^2 hour	watt/meter^2	3.152 480 8
Btu (thermochemical)/inch^2 second	watt/meter^2	1.634 246 2 $\times$ 10^6
calorie (thermochemical)/cm^2 minute	watt/meter^2	6.973 333 3 $\times$ 10^2

Listing by Physical Quantity (*continued*)

To convert from	To	Multiply by
Energy/area time		
erg/centimeter^2 second	watt/meter^2	$1.00^* \times 10^{-3}$
watt/centimeter^2	watt/meter^2	$1.00^* \times 10^4$
Force		
dyne	newton	$1.00^* \times 10^{-5}$
kilogram force (kg$_f$)	newton	$9.806\ 65^*$
kilopond force	newton	$9.806\ 65^*$
kip	newton	$4.448\ 221\ 615\ 260\ 5^* \times 10^3$
lb$_f$ (pound force, avoirdupois)	newton	$4.448\ 221\ 615\ 260\ 5^*$
ounce force (avoirdupois)	newton	$2.780\ 138\ 5 \times 10^{-1}$
pound force (lb$_f$, avoirdupois)	newton	$4.448\ 221\ 615\ 260\ 5^*$
poundal	newton	$1.382\ 549\ 543\ 76^* \times 10^{-1}$
Length		
angstrom	meter	$1.00^* \times 10^{-10}$
astronomical unit (IAU)	meter	$1.496\ 00 \times 10^{11}$
astronomical unit (radio)	meter	$1.495\ 978\ 9 \times 10^{11}$
cable	meter	$2.194\ 56^* \times 10^2$
caliber	meter	$2.54^* \times 10^{-4}$
chain (engineer or ramden)	meter	$3.048^* \times 10^1$
chain (surveyor or gunter)	meter	$2.011\ 68^* \times 10^1$
cubit	meter	$4.572^* \times 10^{-1}$
fathom	meter	1.8288^*
fermi (femtometer)	meter	$1.00^* \times 10^{-15}$
foot	meter	$3.048^* \times 10^{-1}$
foot (U.S. survey)	meter	$1200/3937^*$
foot (U.S. survey)	meter	$3.048\ 006\ 096 \times 10^{-1}$
furlong	meter	$2.011\ 68^* \times 10^2$
hand	meter	$1.016^* \times 10^{-1}$
inch	meter	$2.54^* \times 10^{-2}$
league (international nautical)	meter	$5.556^* \times 10^3$
league (statute)	meter	$4.828\ 032^* \times 10^3$
league (U.K. nautical)	meter	$5.559\ 552^* \times 10^3$
light year	meter	$9.460\ 55 \times 10^{15}$
link (engineer or ramden)	meter	$3.048^* \times 10^{-1}$
link (surveyor or gunter)	meter	$2.011\ 68^* \times 10^{-1}$
meter	wavelengths Kr-86	$1.650\ 763\ 73^* \times 10^6$
micron	meter	$1.00^* \times 10^{-6}$
mil	meter	$2.54^* \times 10^{-5}$
mile (international nautical)	meter	$1.852^* \times 10^3$
mile (U.K. nautical)	meter	$1.853\ 184^* \times 10^3$
mile (U.S. nautical)	meter	$1.852^* \times 10^3$
mile (U.S. statute)	meter	$1.609\ 344^* \times 10^3$
nautical mile (international)	meter	$1.852^* \times 10^3$
nautical mile (U.K.)	meter	$1.853\ 184^* \times 10^3$
nautical mile (U.S.)	meter	$1.852^* \times 10^3$

Listing by Physical Quantity (*continued*)

To convert from	To	Multiply by
Length		
pace	meter	$7.62* \times 10^{-1}$
parsec (IAU)	meter	$3.085\ 7 \times 10^{16}$
perch	meter	$5.0292*$
pica (printer's)	meter	$4.217\ 517\ 6* \times 10^{-3}$
point (printer's)	meter	$3.514\ 598* \times 10^{-4}$
pole	meter	$5.0292*$
rod	meter	$5.0292*$
skein	meter	$1.097\ 28* \times 10^{2}$
span	meter	$2.286* \times 10^{-1}$
statute mile (U.S.)	meter	$1.609\ 344* \times 10^{3}$
yard	meter	$9.144* \times 10^{-1}$
Mass		
carat (metric)	kilogram	$2.00* \times 10^{-4}$
grain	kilogram	$6.479\ 891* \times 10^{-5}$
gram	kilogram	$1.00* \times 10^{-3}$
gram (avoirdupois)	kilogram	$1.771\ 845\ 195\ 312\ 5* \times 10^{-3}$
gram (troy or apothecary)	kilogram	$3.887\ 934\ 6* \times 10^{-3}$
hundredweight (long)	kilogram	$5.080\ 234\ 544* \times 10^{1}$
hundredweight (short)	kilogram	$4.535\ 923\ 7* \times 10^{1}$
kg_f second^2/meter (mass)	kilogram	$9.806\ 65*$
kilogram mass	kilogram	$1.00*$
lb_m (pound mass, avoirdupois)	kilogram	$4.535\ 923\ 7* \times 10^{-1}$
ounce mass (avoirdupois)	kilogram	$2.834\ 952\ 312\ 5* \times 10^{-2}$
ounce mass (troy or apothecary)	kilogram	$3.110\ 347\ 68* \times 10^{-2}$
pennyweight	kilogram	$1.555\ 173\ 84* \times 10^{-3}$
pound mass (lb_m, avoirdupois)	kilogram	$4.535\ 923\ 7* \times 10^{-1}$
pound mass (troy or apothecary)	kilogram	$3.732\ 417\ 216* \times 10^{-1}$
scruple (apothecary)	kilogram	$1.295\ 978\ 2* \times 10^{-3}$
slug	kilogram	$1.459\ 390\ 29 \times 10^{1}$
ton (assay)	kilogram	$2.916\ 666\ 6 \times 10^{-2}$
ton (long)	kilogram	$1.016\ 046\ 908\ 8* \times 10^{3}$
ton (metric)	kilogram	$1.00* \times 10^{3}$
ton (short, 2 000 pound)	kilogram	$9.071\ 847\ 4* \times 10^{2}$
tonne	kilogram	$1.00* \times 10^{3}$
Power		
Btu (thermochemical)/second	watt	$1.054\ 350\ 264\ 488 \times 10^{3}$
Btu (thermochemical)/minute	watt	$1.757\ 250\ 4 \times 10^{1}$
calorie (thermochemical)/second	watt	$4.184*$
calorie (thermochemical)/minute	watt	$6.973\ 333\ 3 \times 10^{-2}$
ft·lb_f/h	watt	$3.766\ 161\ 0 \times 10^{-4}$
ft·lb_f/min	watt	$2.259\ 696\ 6 \times 10^{-2}$
ft·lb_f/s	watt	$1.355\ 817\ 9$
horsepower (550 ft·lb_f/s)	watt	$7.456\ 998\ 7 \times 10^{2}$
horsepower (boiler)	watt	$9.809\ 50 \times 10^{3}$

Listing by Physical Quantity (*continued*)

To convert from	To	Multiply by
Power		
horsepower (electric)	watt	$7.46^* \times 10^2$
horsepower (metric)	watt	$7.354\ 99 \times 10^2$
horsepower (U.K.)	watt	7.457×10^2
horsepower (water)	watt	$7.460\ 43 \times 10^2$
kilocalorie (thermochemical)/minute	watt	$6.973\ 333\ 3 \times 10^1$
kilocalorie (thermochemical)/second	watt	$4.184^* \times 10^3$
watt (international of 1948)	watt	$1.000\ 165$
Pressure		
atmosphere	newton/meter^2	$1.013\ 25^* \times 10^5$
bar	newton/meter^2	$1.00^* \times 10^5$
barye	newton/meter^2	$1.00^* \times 10^{-1}$
centimeter of mercury (0°C)	newton/meter^2	$1.333\ 22 \times 10^3$
centimeter of water (4°C)	newton/meter^2	$9.806\ 38 \times 10^1$
dyne/centimeter^2	newton/meter^2	$1.00^* \times 10^{-1}$
foot of water (39.2°F)	newton/meter^2	$2.988\ 98 \times 10^3$
inch of mercury (32°F)	newton/meter^2	$3.386\ 389 \times 10^3$
inch of mercury (60°F)	newton/meter^2	$3.376\ 85 \times 10^3$
inch of water (39.2°F)	newton/meter^2	$2.490\ 82 \times 10^2$
inch of water (60°F)	newton/meter^2	2.4884×10^2
kg$_f$/cm^2	newton/meter^2	$9.806\ 65^* \times 10^4$
kg$_f$/m^2	newton/meter^2	$9.806\ 65^*$
lb$_f$/ft^2	newton/meter^2	$4.788\ 025\ 8 \times 10^1$
lb$_f$/in.^{2} (psi)	newton/meter^2	$6.894\ 757\ 2 \times 10^3$
millibar	newton/meter^2	$1.00^* \times 10^2$
millimeter of mercury (0°C)	newton/meter^2	$1.333\ 224 \times 10^2$
pascal	newton/meter^2	1.00^*
psi (lb$_f$/in.^{2})	newton/meter^2	$6.894\ 757\ 2 \times 10^3$
torr (0°C)	newton/meter^2	$1.333\ 22 \times 10^2$
Speed		
foot/hour	meter/second	$8.466\ 666\ 6 \times 10^{-5}$
foot/minute	meter/second	$5.08^* \times 10^{-3}$
foot/second	meter/second	$3.048^* \times 10^{-1}$
inch/second	meter/second	$2.54^* \times 10^{-2}$
kilometer/hour	meter/second	$2.777\ 777\ 8 \times 10^{-1}$
knot (international)	meter/second	$5.144\ 444\ 444 \times 10^{-1}$
mile/hour (U.S. statute)	meter/second	$4.4704^* \times 10^{-1}$
mile/minute (U.S. statute)	meter/second	$2.682\ 24^* \times 10^1$
mile/second (U.S. statute)	meter/second	$1.609\ 344^* \times 10^3$
Temperature		
Celsius	kelvin	$t_K = t_C + 273.15$
Fahrenheit	kelvin	$t_K = (5/9)(t_F + 459.67)$
Fahrenheit	Celsius	$t_C = (5/9)(t_F - 32)$
Rankine	kelvin	$t_K = (5/9)t_R$

Listing by Physical Quantity (*continued*)

To convert from	To	Multiply by
Time		
day (mean solar)	second (mean solar)	$8.64^* \times 10^4$
day (sidereal)	second (mean solar)	$8.616\ 409\ 0 \times 10^4$
hour (mean solar)	second (mean solar)	$3.60^* \times 10^3$
hour (sidereal)	second (mean solar)	$3.590\ 170\ 4 \times 10^3$
minute (mean solar)	second (mean solar)	$6.00^* \times 10^1$
minute (sidereal)	second (mean solar)	$5.983\ 617\ 4 \times 10^1$
month (mean calendar)	second (mean solar)	$2.628^* \times 10^6$
second (ephemeris)	second	$1.000\ 000\ 000$
second (mean solar)	second (ephemeris)	Consult American Ephemeris and Nautical Almanac
second (sidereal)	second (mean solar)	$9.972\ 695\ 7 \times 10^{-1}$
year (calendar)	second (mean solar)	$3.1536^* \times 10^7$
year (sidereal)	second (mean solar)	$3.155\ 815\ 0 \times 10^7$
year (tropical)	second (mean solar)	$3.155\ 692\ 6 \times 10^7$
year 1900, tropical, Jan., day 0, hour 12	second (ephemeris)	$3.155\ 692\ 597\ 47^* \times 10^7$
Viscosity		
centipoise	newton second/meter^2	$1.00^* \times 10^{-3}$
centistoke	meter^2/second	$1.00^* \times 10^{-6}$
foot^2/second	meter^2/second	$9.290\ 304^* \times 10^{-2}$
lb$_m$/ft·s	newton second/meter^2	$1.488\ 163\ 9$
lb$_f$·s/ft^2	newton second/meter^2	$4.788\ 025\ 8 \times 10^1$
poise	newton second/meter^2	$1.00^* \times 10^{-1}$
poundal second/foot^2	newton second/meter^2	$1.488\ 163\ 9$
rhe	meter^2/newton second	$1.00^* \times 10^1$
slug/foot second	newton second/meter^2	$4.788\ 025\ 8 \times 10^1$
stoke	meter^2/second	$1.00^* \times 10^{-4}$
Volume		
acre foot	meter^3	$1.233\ 481\ 837\ 547\ 52^* \times 10^3$
barrel (petroleum, 42 gallons)	meter^3	$1.589\ 873 \times 10^{-1}$
board foot	meter^3	$2.359\ 737\ 216^* \times 10^{-3}$
bushel (U.S.)	meter^3	$3.523\ 907\ 016\ 688^* \times 10^{-2}$
cord	meter^3	$3.624\ 556\ 3$
cup	meter^3	$2.365\ 882\ 365^* \times 10^{-4}$
dram (U.S. fluid)	meter^3	$3.696\ 691\ 195\ 312\ 5^* \times 10^{-6}$
fluid ounce (U.S.)	meter^3	$2.957\ 352\ 956\ 25^* \times 10^{-5}$
foot^3	meter^3	$2.831\ 684\ 659\ 2^* \times 10^{-2}$
gallon (U.K. liquid)	meter^3	$4.546\ 087 \times 10^{-3}$
gallon (U.S. dry)	meter^3	$4.404\ 883\ 770\ 86^* \times 10^{-3}$
gallon (U.S. liquid)	meter^3	$3.785\ 411\ 784^* \times 10^{-3}$
gill (U.K.)	meter^3	$1.420\ 652 \times 10^{-4}$
gill (U.S.)	meter^3	$1.182\ 941\ 2 \times 10^{-4}$
hogshead (U.S.)	meter^3	$2.384\ 809\ 423\ 92^* \times 10^{-1}$
inch^3	meter^3	$1.638\ 706\ 4^* \times 10^{-5}$
liter	meter^3	$1.00^* \times 10^{-3}$

Listing by Physical Quantity (*continued*)

To convert from	To	Multiply by
	Volume	
ounce (U.S. fluid)	meter^3	$2.957\ 352\ 956\ 25^* \times 10^{-5}$
peck (U.S.)	meter^3	$8.809\ 767\ 541\ 72^* \times 10^{-3}$
pint (U.S. dry)	meter^3	$5.506\ 104\ 713\ 575^* \times 10^{-4}$
pint (U.S. liquid)	meter^3	$4.731\ 764\ 73^* \times 10^{-4}$
quart (U.S. dry)	meter^3	$1.101\ 220\ 942\ 715^* \times 10^{-3}$
quart (U.S. liquid)	meter^3	$9.463\ 529\ 5 \times 10^{-4}$
stere	meter^3	1.00^*
tablespoon	meter^3	$1.478\ 676\ 478\ 125^* \times 10^{-5}$
teaspoon	meter^3	$4.928\ 921\ 593\ 75^* \times 10^{-6}$
ton (register)	meter^3	$2.831\ 684\ 659\ 2^*$
yard^3	meter^3	$7.645\ 548\ 579\ 84^* \times 10^{-1}$

Physical Constants*

Quantity	Symbol	Value	Error ppm	Prefix	Unit
Speed of light in vacuum	c	2.997 925 0	0.33	$\times 10^8$	$\mathrm{m \cdot s^{-1}}$
Gravitational constant	G	6.673 2	460	10^{-11}	$\mathrm{N \cdot m^2 \cdot kg^{-2}}$
Avogadro constant	N_A	6.022 169	6.6	10^{26}	$\mathrm{kmol^{-1}}$
Boltzmann constant	k	1.380 622	43	10^{-23}	$\mathrm{J \cdot K^{-1}}$
Gas constant	R	8.314 34	42	10^3	$\mathrm{J \cdot kmol^{-1} \cdot K^{-1}}$
Faraday constant	F	9.648 670	5.5	10^7	$\mathrm{C\ kmol^{-1}}$
Unified atomic mass unit	u	1.660 531	6.6	10^{-27}	kg
Planck constant	h	6.626 196	7.6	10^{-34}	$\mathrm{J \cdot s}$
	$h/2\pi$	1.054 591 9	7.6	10^{-34}	$\mathrm{J \cdot s}$
Electron charge	e	1.602 191 7	4.4	10^{-19}	C
Electron rest mass	m_e	9.109 558	6.0	10^{-31}	kg
Proton rest mass	m_p	1.672 614	6.6	10^{-27}	kg
Neutron rest mass	m_n	1.674 920	6.6	10^{-27}	kg
Electron charge to mass ratio	e/m_e	1.758 802 8	3.1	10^{11}	$\mathrm{C \cdot kg^{-1}}$
Stefan-Boltzmann constant	σ	5.669 61	170	10^{-8}	$\mathrm{W \cdot m^{-2} \cdot K^{-4}}$
First radiation constant	$2\pi hc^2$	3.741 844	7.6	10^{-16}	$\mathrm{W \cdot m^2}$
Second radiation constant	hc/k	1.438 833	43	10^{-2}	$\mathrm{m \cdot K}$
Rydberg constant	R^∞	1.097 373 12	0.10	10^7	$\mathrm{m^{-1}}$
Bohr radius	a_0	5.291 771 5	1.5	10^{-11}	m
Classical electron radius	r_e	2.817 939	4.6	10^{-15}	m
Compton wavelength of electron	λ_C	2.426 309 6	3.1	10^{-12}	m
	$\lambda_C/2\pi$	3.861 592	3.1	10^{-13}	m
Compton wavelength of proton	$\lambda_{C,p}$	1.321 440 9	6.8	10^{-15}	m
	$\lambda_{C,p}/2\pi$	2.103 139	6.8	10^{-16}	m
Compton wavelength of neutron	$\lambda_{C,n}$	1.319 621 7	6.8	10^{-15}	m
	$\lambda_{C,n}/2\pi$	2.100 243	6.8	10^{-16}	m
Electron magnetic moment	μ_e	9.284 851	7.0	10^{-24}	$\mathrm{J \cdot T^{-1}}$

Physical Constants* (*continued*)

Quantity	Symbol	Value	Error ppm	Prefix	Unit
Proton magnetic moment	μ_p	1.410 620 3	7.0	10^{-26}	$J \cdot T^{-1}$
Bohr magneton	μ_B	9.274 096	7.0	10^{-24}	$J \cdot T^{-1}$
Nuclear magneton	μ_n	5.050 951	10	10^{-27}	$J \cdot T^{-1}$
Gyromagnetic ratio of	γ'_p	2.675 127 0	3.1	10^8	$rad \cdot s^{-1} \cdot T^{-1}$
protons in H_2O	$\gamma'_p/2\pi$	4.257 597	3.1	10^7	$Hz \cdot T^{-1}$
Gyromagnetic ratio of pro-	γ_p	2.675 196 5	3.1	10^8	$rad \cdot s^{-1} \cdot T^{-1}$
tons in H_2O corrected for	$\gamma_p/2\pi$	4.257 707	3.1	10^7	$Hz \cdot T^{-1}$
diamagnetism of H_2O					
Magnetic flux quantum	Φ_0	2.067 853 8	3.3	10^{-15}	Wb
Quantum of circulation	$h/2m_e$	3.636 947	3.1	10^{-4}	$J \cdot s \cdot kg^{-1}$
	h/m_e	7.273 894	3.1	10^{-4}	$J \cdot s \cdot kg^{-1}$

Reprinted from NASA publication SP 7012, *The International System of Units, Physical Constants and Conversion Factors*, by E. A. Mechtly, 2nd revision, 1973.

Other Important Constants
π = 3.141 592 653 589
e = 2.718 281 828 459
μ_0 = $4\pi \times 10^{-7}$ H/m (exact), permeability of free space
= 1.256 637 061 $\times 10^{-6}$ H/m
ϵ_0 = $\mu_0^{-1}c^{-2}$ F/m, permittivity of free space
= 8.854 185 $\times 10^{-12}$ F/m

Specific Gravities and Average Densities for Common Materials

Gases

Gases	Average density	
	kg/m³	**(lbₘ/ft³)**
Air	1.284	(0.080 18)
Ammonia	0.771 0	(0.048 13)
Carbon dioxide	1.977	(0.123 4)
Carbon monoxide	1.251	(0.078 06)
Ethane	1.357	(0.084 69)
Helium	0.178 4	(0.011 14)
Hydrogen	0.089 9	(0.005 61)
Methane	0.717 6	(0.044 80)
Nitrogen	1.251	(0.078 07)
Oxygen	1.429	(0.089 21)
Sulfur dioxide	2.927	(0.182 7)

Condition: 0°C and standard atmospheric pressure.

For liquids and solids, **specific gravity** is defined to be the ratio of the density of the material to the density of water.

Liquids

Liquids	Average specific gravity	Average density kg/m³	(lb$_m$/ft³)
Alcohol	0.80	800	(50)
Benzene	0.88	880	(55)
Gasoline	0.67	670	(42)
Glycerine	1.25	1 250	(78)
Kerosene	0.80	800	(50)
Oil	0.88	880	(55)
Turpentine	0.87	870	(54)
Water, fresh	1.00	1 000	(62.4)
Water, salt	1.03	1 030	(64.3)

Condition: Standard temperature and pressure.

Coefficients of Friction

Surfaces	Static	Kinetic
Wood/wood	0.3–0.5	0.25–0.4
Wood/glass	≈0.37	≈0.30
Wood/iron	0.43–0.50	0.38–0.45
Wood/metals	0.40–0.60	0.35–0.60
Wood/leather	0.38–0.45	0.30–0.35
Glass/glass	≈0.24	0.20–0.25
Glass/leather	≈0.36	≈0.34
Iron/iron	0.4–0.5	0.4–0.5
Iron/leather	0.45–0.50	0.35–0.40
Iron/brass	0.35–0.45	0.30–0.35
Steel/babbitt	0.35–0.40	0.30–0.35
Steel/ice	0.03–0.04	0.03–0.04

Solids (Nonmetallic)

Material	Average specific gravity	Average density kg/m³	(lb$_m$/ft³)
Asbestos	2.5	2 500	(153)
Brick	1.80	1 800	(112)
Cedar	0.35	350	(22)
Cement, portland	1.44	1 440	(90)
Coal, anthracite	1.60	1 600	(95)
Coal, bituminous	1.35	1 350	(84)
Concrete	2.30	2 300	(144)

Solids (Nonmetallic) (*continued*)

Material	Average specific gravity	Average density kg/m³	(lbₘ/ft³)
Douglas fir	0.50	500	(31)
Glass, common	2.65	2 650	(165)
Granite, solid	2.70	2 700	(172)
Gravel	1.55	1 550	(97)
Gypsum	2.31	2 310	(144)
Leather	0.94	940	(59)
Limestone, solid	2.45	2 450	(153)
Mahogany	0.54	540	(34)
Marble	2.75	2 750	(172)
Oak, white	0.77	770	(48)
Paper	0.93	930	(58)
Pine, white	0.43	430	(27)
Redwood	0.42	420	(26)
Rubber	0.94	940	(59)
Salt	1.00	1 000	(62)
Sand, loose	1.55	1 550	(97)
Sugar	1.61	1 610	(101)
Sulfur	2.1	2 100	(131)

Metals

Metals	Average specific gravity	Average density kg/m³	(lbₘ/ft³)
Aluminum	2.64	2 640	(165)
Brass, cast	8.57	8 570	(535)
Bronze	8.17	8 170	(510)
Copper, cast	8.9	8 900	(555)
Gold	19.3	19 300	(1 210)
Iron, cast	7.05	7 050	(440)
Iron, ore	5.21	5 210	(325)
Iron, wrought	7.77	7 770	(485)
Lead	11.3	11 300	(705)
Manganese	7.4	7 400	(462)
Mercury	13.6	13 600	(849)
Nickel	8.9	8 900	(556)
Silver	10.5	10 500	(655)
Steel, cold drawn	7.83	7 830	(489)
Steel, structural	7.90	7 900	(493)
Steel, tool	7.70	7 700	(481)
Tin, cast	7.30	7 300	(456)
Titanium	4.5	4 500	(281)
Uranium	18.7	18 700	(1 170)
Zinc, cast	7.05	7 050	(440)

Appendix B
Library Reference Materials

DICTIONARIES

Crowley, Ellen T., and Thomas, Robert C. *Trade Name Dictionary*. 1st ed. 2 vols. Detroit: Gale Research, 1976. This dictionary of popular trade names includes model names and design names, as well as brand names and product names of over 100 000 consumer products. The addresses of their manufacturers and distributors are included.

Gardner, William. *Chemical Synonyms and Trade Names*. 8th ed. Oxford, England: Technical Press, 1978. Included are 35 000 chemical definitions with an index. The manufacturers are given for the proprietary chemical names listed.

Illustrated Encyclopedic Dictionary of Building and Construction Terms. Edited by Hugh Brooks. Englewood Cliffs, N.J.: Prentice-Hall, 1976. This dictionary provides definitions for terms in all areas related to construction, including real estate, insurance, mathematics, surveying, and engineering. Additional features include an "index by functions," which lists, alphabetically, all terms in the dictionary in each of twenty-three subject areas to help you with the vocabulary of a specific area. It also includes mathematical charts and tables used in construction and a dictionary of construction-related organizations and trade unions.

Institute of Electrical and Electronics Engineers. *IEEE Standard Dictionary of Electrical and Electronics Terms*. New York: Wiley-Interscience, 1972. Definitions in this dictionary are derived from three major sources: IEEE Standards, American National Standards, and IEC Recommendations. The preferred term appears first followed by variations of the term in descending order of preferred usage. The reader is also given a reference to the document in which the term was originally defined.

Sippl, Charles J., and Kidd, David A. *Microcomputer Dictionary and Guide*. 1st ed. Champagne, Ill.: Matrix Publishers, 1975. Included are over 5 000 entries of terms and definitions along with product, procedure, and application explanations. The definitions are given in terms used by product manufacturers, system designers, and applications developers.

239

Sneddon, I. N. *Encyclopedic Dictionary of Mathematics for Engineers and Applied Scientists*. 1st ed. New York: Pergamon Press, 1976. This dictionary attempts to define the mathematical concepts most frequently used in engineering, with the emphasis on applications, rather than theory, of mathematics. Cross-references lead the user of the dictionary to the section where the term and related terms are discussed.

TECHNICAL ENCYCLOPEDIAS

Belzer, Jack; Holzman, Albert G.; Kent, Allen, eds. *Encyclopedia of Computer Science and Technology*. Vols. 1–14. New York: Dekker, 1975–80. Still in progress, this encyclopedia contains articles written by experts in computer and information science. Volume 14 was published in 1980 and contains entries from "Very Large Data Base System" to "Zero-memory and Markov Information Source."

Kirk, R. E., and Othmer, D. F., eds. *Encyclopedia of Chemical Technology*. 2nd ed. 22 vols. New York: Wiley, 1963–70.

Kirk, R. E., and Othmer, D. F., eds. *Encyclopedia of Chemical Technology*. 3rd ed. New York: Wiley, 1979. Volumes in progress. The third edition of this encyclopedia, which discusses industrial chemical processes, supplements the second edition. The index is particularly helpful for looking up chemical processes that have been given their originator's name, or for locating discussions of a chemical's use in different procedures.

Lapedes, Daniel N. *McGraw-Hill Encyclopedia of Energy*. 2nd ed. New York: McGraw-Hill, 1980. This one-volume encyclopedia contains over 300 articles written by specialists. It is divided into two sections. The first, "Energy Perspectives," contains articles concerning energy choices and planning for the future. The second is "Energy Technology," in which articles discuss technical subjects such as nuclear power, fusion, and wind power.

Mark, Herman F.; Gaylor, Norman G.; and Bikales, Robert M. *Encyclopedia of Polymer Science and Technology: Plastics, Resin, Rubbers, Fibers*. 16 vols. New York: Interscience Publishers, 1964–72. This encyclopedia covers both the theory and the applications of this important field. The articles are written for the reader who is already familiar with the basics of chemistry, physics, or engineering.

McGraw-Hill Encyclopedia of Science and Technology. 4th ed. 15 vols. plus yearly supplements. New York: McGraw-Hill. A very comprehensive encyclopedia covering most areas of science and technology.

HANDBOOKS

General Engineering

Bolz, Ray E. *CRC Handbook of Tables for Applied Engineering Science*. Cleveland: CRC Press, 1973. Includes data in all areas of engineering for the practicing engineer. Data are given in SI units as well as conventional units.

Eshbach, Ovid W., and Sonders, Mott. *Handbook of Engineering Funda-mentals*. 3rd ed. New York: Wiley, 1975. Contains data in mathematics, mechanics, engineering thermodynamics, electromagnetics and electronics, radiation, lights, acoustics, heat transfer, automatic control, chemistry, and properties of materials.

Perry, Robert H. *Engineering Manual*. 3rd ed. New York: McGraw-Hill, 1976. Includes design methods; data for building construction as well as all areas of engineering.

Chemical Engineering

Perry, John H., and Chilton, Cecil H., eds. *Chemical Engineer's Handbook*. 5th ed. New York: McGraw-Hill, 1973. Covers basic topics in chemical en-gineering, including heat transfer, refrigeration, and process control.

Civil Engineering

Merritt, Frederick S., ed. *Standard Handbook for Civil Engineers*. 2nd ed. New York: McGraw-Hill, 1976. Discusses computer use in engineering and discusses specifications, construction, surveying, and municipal and regional planning.

Electrical Engineering

Garland, J. D. *National Electrical Code Reference Book: Based on the 1978 Code*. 2nd ed. Englewood Cliffs, N.J.: Prentice-Hall, 1979. Interpretations and clarifications of the national electrical code are included in this handbook, along with definitions and diagrams.

Fink, Donald G., ed. *Standard Handbook for Electrical Engineering*. 11th ed. New York: McGraw-Hill, 1978. Discusses all aspects of electrical engineering, including generators, power plants, motors, data processing, and power trans-mission.

Environmental Engineering

Bond, Richard G., and Straub, Conrad R. *CRC Handbook of Environmental Control*. 4 vols. Cleveland: CRC Press, 1973–74. Each volume of this handbook covers a separate environmental topic. Volume 1 covers air pollution; Volume 2, solid waste; Volume 3, waste supply and treatment; Volume 4 discusses waste water treatment and disposal.

Industrial Engineering

Lewis, Bernard T., and Marron, J. P., eds. *Facilities and Plant Engineering Handbook*. New York: McGraw-Hill, 1974. Discusses management of facilities and plants. Included are budgeting, personnel administration, data processing, and organization. Other types of engineering are discussed in relation to plan-ning, design, and maintenance of facilities.

Ireson, William Grant, and Grant, E. L., eds. *Handbook of Industrial En-gineering and Management*. 2nd ed. Englewood Cliffs, N.J.: Prentice-Hall, 1971. Chapters in this handbook cover various subjects related to industrial engineering and management. Included are managerial economics, industrial safety, and attitudes of labor toward industrial engineering methods.

Materials Engineering

American Society for Metals. *Metals Handbook*. 11 vols. Metals Park, Ohio: The Society, 1961–76. Volumes of this handbook cover properties of metals, selection of metals, machining, forming, casting, welding, and brazing. Also covered are alloys, metallography, failure analysis and prevention, and non-destructive testing.

Lynch, Charles T. *CRC Handbook of Materials Science*. 3 vols. Cleveland: CRC Press, 1974–75. This handbook gives data on the properties of metals, oxides, glasses, polymers, composites, electronic materials, and nuclear materials.

Society of the Plastics Industry. *Plastics Engineering Handbook*. 4th ed. Edited by Joel Frados. New York: Van Nostrand Reinhold, 1976. The properties, process methods, and performance data of plastics are covered in this standard reference source.

Mechanical Engineering

Baumeister, T., and Marks, L., eds. *Standard Handbook for Mechanical Engineers*. 8th ed. New York: McGraw-Hill, 1978. A comprehensive handbook covering subjects such as materials, mechanical properties, mechanical phenomena, and transportation.

Kent's Mechanical Engineers' Handbook. 12th ed. 2 vols. Volume 1, "Power," edited by J. K. Salisbury. Volume 2, "Design and Production," edited by C. Carmichael. New York: Wiley, 1950. Directed toward engineers who design and manufacture products or engineers who deal with thermal combustion, heat transfer, transportation, and so on.

Nuclear Engineering

U.S. Atomic Energy Commission. *Nuclear Power Reactor Instrumentation Systems Handbook*. Washington, D.C.: U.S. Government Printing Office, 1973.

U.S. Atomic Energy Commission. *Reactor Handbook*. 2nd ed. 5 vols. New York: Wiley, 1960–64. Volume 1, "Materials"; Volume 2, "Fuel Processing"; Volume 3, Part A, "Nuclear Physics"; Volume 3, Part B, "Shielding"; Volume 4, "Nuclear Engineering."

INDEXES AND ABSTRACTS

General Science and Engineering

Applied Science and Technology Index. New York: H. W. Wilson, 1913 to the present. Monthly. This index covers a selected list of frequently used English language journals in engineering and physical science. It is a subject index only; no author index is included. It looks very much like *Reader's Guide to Periodical Literature*, an index to news magazines, which is also published by the H. W. Wilson Company. It contains no abstracts.

Engineering Index. New York: Engineering Index, 1884 to the present. Monthly. This is the most comprehensive index to general engineering literature. It covers over 2 000 journals as well as proceedings and technical reports of engineering societies, universities, laboratories, and government agencies. It is a subject index with abstracts; each issue includes an author index. It is computer searchable in a magnetic tape version called COMPENDEX.

Science Citation Index. Philadelphia: Institute for Scientific Information, 1961 to the present. Quarterly. Indexes current articles and the footnotes cited in those articles for over 2 500 of the world's scientific and technical journals. It consists of three indexes: the *Source Index,* an index to the authors of the articles covered; the *Citation Index,* an index to the footnotes of the articles in the *Source Index;* and the *Permuterm Subject Index,* a permuted index of the words contained in the titles of the source articles. In combination with the *Citation Index,* one can find newer journal articles that reference a given older article.

Aerospace Engineering

International Aerospace Abstracts. New York: American Institute of Aeronautics and Astronautics, 1961 to the present. This index, with abstracts, covers periodicals, books, and conference papers in aeronautics and space technology. It covers the periodical literature as a companion volume to *Scientific and Technical Aerospace Reports,* which covers aerospace report literature.

Chemical Engineering

Chemical Abstracts. Columbus, Ohio: American Chemical Society, 1907 to the present. Weekly. Covers all technical literature related to chemistry and chemical engineering. It has chemical names and formula indexes, as well as subject and author indexes. It is computer searchable via the data base CHEMCON.

Civil Engineering

ASCE Publications Abstracts. New York: American Society of Civil Engineers, 1966 to the present. Bimonthly. This index includes abstracts of all papers in ASCE journals.

Computer Engineering

Computer and Control Abstracts. London: Institution of Electrical and Electronics Engineers; New York: Institute of Electrical and Electronics Engineers, 1966 to the present. Monthly. This index, which is section C of *Science Abstracts,* covers journals, conference papers, patents, and books on computers. It also is computer searchable as part of the data base INSPEC.

Computer and Information Systems. Riverdale, Md.: Cambridge Scientific Abstracts, 1962 to the present. This index covers all primary sources (journals, books, technical reports, patents) on computer software and applications.

Electrical Engineering

Electrical and Electronics Abstracts. London: Institution of Electrical and Engineers; New York: Institute of Electrical and Electronics Engineers, 1966 to the present. Monthly. An index with abstracts, also known as Part B of *Science*

Abstracts, this work covers over 2 000 journals as well as books, papers, patents, and reports. It is computer searchable as part of the INSPEC data base.

Electronics and Communications Abstracts Journal. Riverdale, Md.: Cambridge Scientific Abstracts, 1967 to the present. 10/year. This index covers periodicals, government reports, conference proceedings, books, and patents.

Index to IEEE publications. New York: Institute of Electrical and Electronics Engineers, 1971 to the present. Annual. This work is an annual index to publications of the Institute of Electrical and Electronics Engineers, including the IEEE Transactions.

ENERGY ENGINEERING

Energy Index. New York: Environment Information Center, 1973 to the present. Bimonthly. This abstracting service covers more than 1 000 scientific and trade journals, as well as reports, surveys, and documents.

Petroleum Abstracts. Tulsa, Okla.: University of Tulsa, 1961 to the present. Weekly. Coverage includes all aspects of the petroleum industry: geology, engineering, chemistry.

Environmental Engineering

Environment Index. New York: Environment Information Center, 1971 to the present. Annual.

Environment Abstracts. New York: Environment Information Center, 1974 to the present. Monthly. A monthly abstracting service covering material from scientific and technical periodicals, conferences, and technical reports. *Environment Index* is the annual index to these abstracts.

Environmental Periodicals Bibliography. Santa Barbara, Calif.: Environmental Studies Institute, 1972 to the present. Bimonthly. This index reproduces the table of contents of about 300 environment-related periodicals. It includes a subject index.

Mechanical Engineering

Applied Mechanics Reviews. New York: American Society of Mechanical Engineers, 1948 to the present. Monthly. Arranged by a broad subject classification, with a more detailed key-word subject index. It covers 400 periodicals in applied mechanics.

Metallurgical Engineering

Metals Abstracts. Metals Park, Ohio: American Society for Metals and Metal Society (joint publishers). This index combines and replaces two previous indexes, *Metallurgical Abstracts* (Institute of Metals) and *Review of Metal Literature* (American Society for Metals). It is searchable on the data base METADEX.

Nuclear Engineering

INIS Atomindex. Vienna: International Nuclear Information System, 1970 to the present. Semimonthly. This index is produced by the International Nuclear Information System. Abstracts are provided for books, patents, technical reports, and journal articles. In 1976 it took over the coverage of ERDA (now Department of Energy) reports in nuclear science and superseded *Nuclear Science Abstracts.*

TECHNICAL REPORTS

Government Reports Announcements and Index. Springfield, Va.: U.S. National Technical Information Service, 1946 to present. This semimonthly index covers all types of technical reports prepared by and for most government agencies, including the Department of Defense. The index includes some NASA and ERDA reports that may be covered by other indexes. The technical reports are usually recognizable by series numbers beginning with the letters PB and AD. In addition to covering technical reports, this index covers a wide range of reference tools. (Computer searchable 1964 to the present.)

INIS Atomindex. Vienna: International Atomic Energy Agency, 1970 to the present. A semimonthly index, produced by the International Nuclear Information System (INIS), covering reports related to nuclear energy. This index superseded *Nuclear Science Abstracts* for the coverage of ERDA reports in the field of nuclear energy. (Computer searchable 1976 to the present.)

Monthly Catalog of United States Government Publications. Washington, D.C.: Superintendent of Documents, 1895 to the present. Covers congressional reports and reports from various government agencies. Unlike any of the other technical report indexes listed here, the *Monthly Catalog* is arranged by issuing government agency and does not provide abstracts. While the *Monthly Catalog* has author, title, and subject indexes, it does not have agency, report number, or contract number indexes. (Computer searchable 1976 to the present.)

Scientific and Technical Aerospace Reports (STAR). College Park, Md.: U.S. National Aeronautics and Space Administration, 1963 to the present. This semimonthly index covers the field of aeronautics and astronautics. Unlike the other indexes mentioned, STAR covers only technical reports and not the published literature.

RAND Abstracts. Santa Monica, Calif.: Rand Corporation, 1946 to the present. A quarterly annotated index to Rand Corporation classified reports.

Selected
Bibliography

Adams, H. F. R. *SI Metric Units, An Introduction*. New York: McGraw-Hill Ryerson, 1974.

Advanced Calculator Logic—HP RPN/Algebraic—A Comparative Analysis. Published by Hewlett-Packard Company (#5953-1930).

Aeronautics and Astronautics. Published periodically by the American Institute of Aeronautics and Astronautics (AIAA).

Agricultural Engineering. Published periodically by the American Society of Agricultural Engineers (ASAE).

Angel, J. L. *Directory of Professional and Occupational Licensing in the United States*. New York: World Trade Academy Press, 1970.

Asimov, M. *Introduction to Design*. Englewood Cliffs, N.J.: Prentice-Hall, 1962.

ASTM. *Value Engineering in Manufacturing*. American Society of Tool and Manufacturing Engineers. Englewood Cliffs, N.J.: Prentice-Hall, 1967.

Automotive Engineering. Published periodically by the Society of Automotive Engineers (SAE).

Aviation Space. Published periodically by Aerospace Education.

Ball, John A. *Algorithms for RPN Calculators*. New York: Wiley, 1978.

Beakley, G. C., and Leach, N. W. *Engineering, An Introduction to a Creative Profession*. 3rd ed. New York: Macmillan, 1977.

Bell Telephone Magazine. Published periodically by The Bell Telephone Company.

Blair, R. N., and Whitson, C. W. *Elements of Industrial Systems Engineering*. Englewood Cliffs, N.J.: Prentice-Hall, 1971.

Blanchard, B. S. *Engineering Organization and Management*. Englewood Cliffs, N.J.: Prentice-Hall, 1976.

Burge, D. A. *Patent and Trademark Tactics and Practice*. New York: Wiley, 1980.

Calculator Decision-Making Sourcebook. The Staff of the Texas Instruments Learning Center, Texas Instruments, Inc., 1981.

Chemical Engineering Education. Published periodically by the Chemical Engineering division of the American Society for Engineering Education (ASEE).

Civil Engineering. Published periodically by the American Society of Civil Engineers (ASCE).

Clough, R. H., and Sears, G. A. *Construction Project Management*. 2nd ed. New York: Wiley, 1979.

Cohen, B. L. *Nuclear Science and Society*. Garden City, N.Y.: Doubleday, 1974.

Constance, J. D. *How to Become a Professional Engineer*. 3rd ed. New York: McGraw-Hill, 1978.

Design Engineering. Published periodically by Maclean-Hunter, Toronto.

Dilworth, J. B. *Production and Operations Management*. New York: Random House, 1979.

Eide, A. R.; Jensen, R. D.; Mashaw, L. H.; and Northup, L. L. *Engineering Fundamentals and Problem Solving*. New York: McGraw-Hill, 1979.

Electrical World. Published periodically by McGraw-Hill.

Engineering Education. Published periodically by the ASEE.

Farkas, L. L. *Management of Technical Field Operations*. New York: McGraw-Hill, 1970.

Fletcher, L. S., and Shoup, T. E. *Introduction to Engineering Including FORTRAN Programming*. Englewood Cliffs, N.J.: Prentice-Hall, 1978.

Gibbs-Smith, C. *The Inventions of Leonardo Da Vinci*. Oxford, England: Phaidon Press, 1978.

Glorioso, Robert M., and Hill, Francis S. *Introduction to Engineering*. Englewood Cliffs, N.J.: Prentice-Hall, 1975.

Hajek, V. G. *Management of Engineering Projects*. New York: McGraw-Hill, 1977.

The Hewlett-Packard Personal Calculator Digest. Vol. 7, 1980. Published by the Hewlett-Packard Company.

Hoover, Herbert. *Memoirs of Herbert Hoover*. Vol. 1: *Years of Adventure*. New York: Macmillan, 1951.

IEEE Spectrum. Published periodically by the Institute of Electrical and Electronics Engineers.

Kemper, J. D. *The Engineer and His Profession*. 2nd ed. New York: Holt, Rinehart, & Winston, 1975.

Konzo, S., and Bayne, J. W. *Opportunities in Mechanical Engineering*. Skokie, Ill.: National Textbook Company, 1978.

Krick, E. *An Introduction to Engineering: Concepts, Methods and Issues*. New York: Wiley, 1976.

Materials Engineering. Published periodically by Penton/IPC.

Mechanical Engineering. Published periodically by the American Society of Mechanical Engineers (ASME).

Nonhebel, G. *Chemical Engineering in Practice*. London: Wykeham Publications, 1973.

Professional Engineer. Published periodically by the National Society of Professional Engineers.

Red, W. E. *Engineering: The Career and the Profession*. Monterey, Calif.: Brooks/Cole, 1982.

Red, W. E., and Mooring, B. W. *Engineering: Fundamentals of Problem Solving*. Monterey, Calif.: Brooks/Cole, 1983.

Shoup, Terry E.; Fletcher, Leroy S.; and Mochel, Edward V. *Introduction to Engineering Design with Graphics and Design Projects*. Englewood Cliffs, N.J.: Prentice-Hall, 1981.

Smith, R. J. *Engineering as a Career,* 3rd ed. New York: McGraw-Hill, 1969.

Stein, R. *The Great Inventions*. New York: Ridge Press, 1976.

Transactions of the IEEE. Published periodically by the Institute of Electrical and Electronics Engineers.

Vaughn, R. C. *Introduction to Industrial Engineering*. 2nd ed. Ames, Iowa: Iowa State University Press, 1977.

Whinnery, J. R. *The World of Engineering*. New York: McGraw-Hill, 1965.

Woodson, T. T. *Introduction to Engineering Design*. New York: McGraw-Hill, 1966.

Answers to Problems–
Chapters 5 and 6

Chapter 5

5.1 1.95×10^5

5.2 2.32×10^3

5.3 1.06×10^{-1}

5.4 2.92×10^4

5.5 4.34×10^{-1}

5.6 1.36×10^{15}

5.7 5.03×10^{-1}

5.8 2.35×10^{-8}

5.9 1.15×10^{21}

5.10 6.72×10^4

5.11 2.84×10^3

5.12 (a) 3.96×10^{-2} (c) 9.98×10^{-1}
(b) 7.91×10^{-1} (d) 5.07×10^{-1}

5.13 -2.49×10^{-1}

5.14 9.20

5.15 2.11×10^1

5.16 -3.36

5.17 (a) 1.12×10^{-1} (c) 3.37×10^{-7}
(b) 3.20 (d) 5.68×10^1

5.18 9.98×10^{-1}

5.19 -8.12×10^{-1}

5.20 -2.28

5.21 -1.39×10^{-1}

5.22 -9.21×10^{-1}

5.23 1.00

5.24 1.02×10^2 (deg)

5.25 -1.56 (rad)

5.26 6.09×10^{-1}

5.27 2.49

5.28 7.86×10^{-1}

5.29 2.19×10^1

5.30 $x = -3.61 \times 10^2; y = 6.01 \times 10^2$

5.31 $x = 2.05 \times 10^{-2}; y = 1.49 \times 10^{-2}$

5.32 $r = 1.97 \times 10^{-1}; \theta = -8.10$ (deg)

5.33 $r = 2.18 \times 10^2; \theta = 2.72$ (rad)

5.34 4.02×10^{11}

5.35 1.90×10^4

5.36 5.74

5.37 1.39×10^{-1}

5.38 1.95×10^{-17}

5.39 6.35×10^{22}

5.40 1.36×10^{-1}

5.41 -9.75×10^{-9}

5.42 -3.50×10^1

5.43 5.72×10^{14}

5.44 3.55×10^{-15}

5.45 1.86×10^3

Chapter 6

6.1 False

6.2 π; 1 000 g/kg

6.3 False

6.4 Charge ~ coulomb
Energy ~ kilowatt-hour
Power ~ watt
Current ~ A

Chapter 6 (*continued*)

Velocity $\sim$ km/s
Pressure $\sim$ Pa
Length $\sim$ cubit

6.5 False

6.6 (a) energy, CGS   (d) 57.3
(b) 10^5 (e) 3.28×10^{-2}
(c) J/s   (f) pressure

6.7 a, c, f, g, i (SI)

6.8 SI: M, L, t, T, I, n, I_ℓ
CGS: M, L, t, T, Q, n, I_ℓ
English and engineering:
$\quad F, L, t, T, Q, n, I_\ell$

6.9 Defined units in absolute unit systems are independent of gravitational effects. CGS and SI are absolute unit systems.

6.10 On earth, the magnitude of the mass is the same as the magnitude of the weight. Confusion occurs because people incorrectly equate lb_m to lb_f dimensionally.

6.11 120 kilopascals; 290 K; 1 373 K; 100 kilopascals (to be consistent)

6.12 (b) newtons per meter squared
(c) seconds squared
(d) $3.123\ 6 \times 10^6$ s
(f) 0.301 kg

6.13 (a) $M \cdot L \cdot t^{-2}$ or $M/(L \cdot t^2)$
(b) $L \cdot t^{-1}$ or L/t
(c) $M \cdot L^2 \cdot t^{-3}$ or $M \cdot L^2/t^3$
(d) $M \cdot L^2 \cdot t^{-2}$ or $M \cdot L^2/t^2$
(e) $M \cdot L^{-1} \cdot t^{-2}$ or $M/(L \cdot t^2)$
(f) $L \cdot t^{-2}$ or L/t^2
(g) $M \cdot L^{-3}$ or M/L^3
(h) L^3
(i) $M \cdot L^2 \cdot t^{-2}$ or $M \cdot L^2/t^2$

6.14 a, b

6.15 a, b, c

6.16 $V \sim M \cdot L^2 \cdot t^{-2}$

6.17 $s \sim L$

6.18 $F \sim L, A \sim L \cdot t^{-1}$

6.19 $\dot{m} \sim M \cdot t^{-1}$

6.20 $\omega \sim t^{-1}$

6.21 C_D dimensionless

6.22 $C \sim \sqrt{B}$

6.23 (a) 6.84 km/s
(b) 359° R
(c) $7.867\ 2 \times 10$ dynes
(d) 0.097 4 MPa
(e) 287 rpm
(f) 7.46 kW
(g) 27.6 K
(h) 7.00×10^3 lb_f
(i) 4.2×10^3 kg/m^3
(j) 1.4×10^4 $lb_m \cdot in.^2$
(k) 1.44×10^7 g
(l) 2.24×10^{-3} $ft \cdot lb_f$

6.24 (a) 6.67×10^{-11} $m^3 \cdot kg^{-1} \cdot s^{-2}$
(b) 3.44×10^{-8} $ft^3 \cdot slug^{-1} \cdot s^{-2}$
(c) 1.07×10^{-9} $ft^3 \cdot lb_m^{-1} \cdot s^{-2}$

6.25 5.78×10^6 N

6.26 $q = 8.18 \times 10^2$ Btu/h
Cost = \$0.19 (100% efficiency)
$\quad\quad\;\;$ = \$0.63 (30% efficiency)

6.27 368 N

6.28 1.71×10^3 Btu

6.29 0.155 hp

6.30 12.23 m

6.31 1×10^{-3} kW

6.32 16.6 in.

6.33 1.5×10^{-27} g

6.34 2.37×10^3 lb_f

6.35 5.24×10^{13} W

6.36 100 Ω

6.37 (a) 6.18×10^{16} slugs
(b) 1.99×10^{18} lb_m
(c) 9.03×10^{17} kg

6.38 0.96, dimensionless

6.39 $F_e = 3.56 \times 10^{22}$ N
Speed = 2.98×10^4 m/s
Period = 365.74 d

Index

abstracts, 199
accountability:
  fiscal, 78
  societal, 78
Accreditation Board for Engineering
  and Technology (ABET), 3
addition, 153
aeronautics, 6
aerospace, 30
  technology, 25
aerospace engineering, 5, 6–8, 14
agricultural engineering, 5, 9–10
Alaskan Pipeline, 34
algebraic operating system (AOS),
  144, 153–156
alloying, 27
alphanumerics, 145
American Institute of Chemical Engineers (AIChE), 123
American Institute of Mining, Metallurgical, and Petroleum Engineers
  (AIME), 123
American Society of Civil Engineers
  (ASCE), 123
American Society of Mechanical Engineers (ASME), 123
ampere, 176
Ampere's Law, 173
angle mode, 158, 159
Apollo, 7, 53
applied mechanics, 28
arc melting, 27
architectural engineering, 5
astronautics, 6
avoirdupois, 165

bachelor's degree in engineering, 50
BASIC, 19, 143
bibliographies, 198
biomechanics, 15
biomedical engineering, 5
blasting, 31
Boeing supersonic jet transport (SST),
  43
boring, 31

calculator:
  algebraic operating system, 144
  Casio, 142
  common errors, 158–159
  features, 144–148
  functions, 148–153
  hand-held, 141
  Hewlett-Packard (HP), 142
  LCD displays, 145
  LED displays, 145
  logic, 144
  programmable, 141
  questions about, 141–144
  Radio-Shack, 143
  reversed Polish notation (RPN),
    144, 156–158
  Sharp 143, 144
  Texas Instruments (TI), 143, 148
calculator features:
  key, 144
  magnetic card, 145
  memory, 147
  modules, 147
candela, 178
card readers, 17

catalogs, 198
celsius, 177
centimeter-gram-second unit system
  (CGS), 183
charge, 171
  electric, 180
chemical engineering, 5, 10–13
civil engineering, 5, 13–16
coefficient of performance (COP), 58
communications, 21
computer-aided design (CAD), 19, 63
computer-aided engineering (CAE), 19
computer-aided manufacturing (CAM),
  19
computer engineering, 5, 16–19
computers:
  digital, 16
  hardware, 17
  languages for, 19
  software for, 17–18
  types of, 17–19
compressors, 30
Concorde, 43
construction, 14, 67
construction engineer, 67
  qualifications of, 70
consulting engineer, 51
controls, 28–29
conversion factor, 167
copyrights, 212
Coulomb's Law, 173
craftsperson, 50
cubit, 163
current, 171
  electric, 176

data processing, 24
design, 29, 39, 51, 64
  computer-aided (CAD), 19, 63
  description, 61
  detailed, 47
  education, 41–42
  engineering, 64
  function of, 60
  iterations, 39
  morphology, 42
  negative decisions, 43
  phases of, 42, 45–58
  preliminary, 47
  primary, 39, 64
  problems, 42
  process of, 42, 44–45, 60
  selection, 60
  specification, 60
designer:
  architectural, 61
  characteristics of, 63
  mechanical, 61
  qualifications, 64
development, 51, 64
dictionaries, technical, 197
dimensional analysis, 167
dimensionless quantities, 167
dimensions, 166
  base, 174
  derived, 166, 171
  fundamental, 166
  primary, 166
  reduction of, 170
division, 153
doctor of philosophy (PhD), 48, 104
Dollfus, Charles, 7
drilling, 34
dynamics, 90

education, engineering, 87–104,
    118–119
  cooperative, 102, 106
  engineer-in-training (EIT), 120
  exams and grades, 88, 95, 97
  goals, 89, 90
  homework, 95, 97
  lectures and laboratories, 94–95
  organization and planning, 89–90
  production, 65
electric power and processing, 10
electrical engineering, 5, 20–22
electronics, 21
employer, expectations of, 112
encyclopedias, technical, 198–199

energy/power engineering, 5
energy production, 12, 14, 29
engineering:
  aerospace 5, 6–8, 14
  agricultural, 5, 9–10
  biomedical, 5
  branches of, 4–6
  chemical, 5, 10–13
  civil, 5, 13–16
  computer, 5, 16–19
  computer-aided (CAE), 19
  environmental, 5
  food, 10–11
  forest, 10
  functions, 50–52
  geological, 5
  industrial, 5, 23–24
  marine/naval, 5
  materials and metallurgical, 5,
    25–27
  mechanical, 5, 27–31
  mining, 5, 31–32
  nuclear, 5, 32–33
  petroleum, 5, 33–35
  product, 50
  project, 50
  sales, 73
  societies, 123
  teachers, 51
  technology, 35–36
  testing, 51
engineering sciences, 5
engineers:
  characteristics of, 87
  licensing of, 121–122
  professional, 121
  in training (EIT), 120
Engineer's Council for Professional
    Development (ECPD), 3
engines, 29
English system of units, 164, 171
environmental control, 11
environmental engineering, 5
ethics, engineering:
  ABET Code of, 123–124
  cases in ethical studies, 124–136
equations:
  dimensionally consistent, 167
  energy, 179
  homogeneous, 167
exams, 88, 95, 97
  preparation for, 98–99
exponentiation, 153
extracting, 34

fathoms, 163
feasibility study, 46–47
fiberglass, 54
fieldwork, 69
fluids, 30
food engineering, 10, 11
force, 171, 179
forest engineering, 10
FORTRAN, 19, 143
freon, 56

geiger counter, 31
generators:
  fuel cell, 22
  solar, 22
  wind, 22
geological engineering, 5
geotechnology, 14
germanium, 26
grades, educational, 88
graphics, interactive, 19
graphics terminals, 17, 19
gravitational unit systems, 183–186
Greeks, 163
gross national product (GNP), 67

handbooks, 199
heat:
  exchangers, 30
  pump, 55–58
  transfer, 30
  treatment, 27
heating, ventilating and air condi-
    tioning (HVAC), 30
Hewlett Packard, 156
hierarchy, 151
  automatic, 156
  mistakes, 158, 159
homework, 95
  guidelines, 97
Hoover, Herbert, 117
hydraulics, 31
hydraulics and water resources, 15

industrial engineering, 5, 23–24
Industrial Revolution, 4
Institute of Electrical and Electronics
    Engineers (IEEE), 123
instrumentation, 21–22
international system of units (SI), 165,
    171, 174–183
  derived dimensions of, 178–180
  evolution of, 165
  exceptions to, 182–183

special rules, 180–181
symbols and abbreviations, 181–182
interview, job, 107–108

job:
    decision, 110
    interview, 107–108

kelvin, 176
kilogram, 175, 185
King Edward II, 164
knowledge, 118–119

laboratories, 94–95
learning, guidelines, 92
lectures, 94–95
LED/LCD, 145
length, 164, 175
    units of, 164
library, technical:
    card catalog, 194–195
    circulation, 194
    Dewey Decimal System, 194
    government documents, 204–206
    interlibrary services, 196
    journals, 195–196
    Library of Congress system, 194
    organization of, 194
    reference searches, 207–211
    reference section, 197
    types of, 193
lubrication, 30
Lukasiewicz, Jan, 156
luminous intensity, 178

Magna Carta, 164
magnetic card, 145
magnetic tape, 17
magnetometers, airborne, 31
maintenance, 72
management, 51, 75–77
    people, 76–77
    product and project, 77
management engineering, 23–24
manufacturing engineering, 24
mapping, aerial, 31
marine/naval engineering, 5
mass, 171, 174, 175
master of business administration
    (MBA), 48, 85, 104
master's degree in engineering, 48,
    104
materials/metallurgy engineering, 5,
    25–27

mathematical operations, 182
measurement standard, 175
mechanical engineering, 5, 27–31
memory, 147
metallurgists, 31
metals:
    materials, 30
    multiphase, 27
microcircuitry, 21
microcomputers, 141
microelectronics, 17
microprocessors, 17, 21
mile, meridian, 163
mine/mining:
    filling, 31
    management, 31
    ocean, 31
    surface, 31
    surveying, 31
    timbering, 31
    underground, 31
    ventilation, 31
mining engineering, 5, 31–32
Minuteman Missile, 7
modules, 147
mole, 178
multiplication, 153

National Council of Engineering Examiners (NCEE), 120–121
National Society of Professional Engineers (NSPE), 124
newton, 179
Newton's Law of Gravitation, 174
Newton's Second Law, 173, 179, 183, 185
nuclear engineering, 5, 32–33

on-line bibliographic data bases,
    199–201
operations and maintenance, 51, 70–71
operations engineer, 70

PASCAL, 19
pascal, 180
patents, 211–222
    design, 212
    drama of, 214–222
    plant, 212
    searching, 212–213
    structure 213, 214
    utility, 212
Peace Corps, engineering in, 104

petroleum engineering, 5, 33–35
plane angle, 178
plant, design, 13, 24
plotters, 17, 19
Polish notation, 156
polymer science technology, 13
polymerization, 13
potential difference, 180
pound-force, 183
power and machinery, 10
    generation and transmission, 22
power plants:
    coal-fired, 22
    hydroelectric dams, 22
    nuclear, 22
pressure, 180
pressure vessels, 30
printers, 17, 19
problem solving, 41–42
process definition, 65
process dynamics and control, 11
processes, 52
production and construction, 51, 65
production and quality control, 24
production consumption, 64
production engineering, 34, 65
production engineers, qualifications of,
    66–67
products, 52
professions/professionalism, 101,
    117–118
    characteristics of, 117–118
    competence, 122
    licensing, 120
    loyalty, 122
    responsibilities, 117–118
    service, 122
project engineer, 75

Quadra-Key, 43
quality assurance, 71

radian, 178
real time, 17
reports, 95
research, 51, 52–58, 64
    applied, 52
    basic, 52
    and development (R&D), 54
research engineer, 52, 54
    qualifications, 54
résumé, 105–106
Reverse Polish Notation (RPN), 144,
    156–158

sales, 51, 73
sales engineer, qualifications, 74
sales engineering, 73
sanitation, 14
satellites, earth resources, 31
second, 176
semiconductors, 26
silicon, 26
societies:
  engineering, 123
  technical, 103
software, 19, 63, 148
soils and water, 9
solid angle, 178
solid-state circuitry, 21
space shuttle, 8, 53
speed, 167
SST, 7
standards and specifications, 203–204
statics, 90
steradian, 178
Steven, Simon, 165
structures and environment, 9, 15
subsidence, 31
subtraction, 153
systems analysis, 24

systems engineering, 5

teaching, 51
technical reports, 206–207
technician, engineering, 50
technological team, 39, 48
  members of, 49
temperature, 176
  thermodynamic, 177
testing, 51–52, 59
  field, 57
thermocouples, 30
thermodynamics, 30
time, 176
trademark, 212
transportation, 30

unit operations, 13
units:
  absolute, 174
  alternative, 166
  American engineering, 166, 174
  base, 175–180
  centimeter-gram-second (CGS), 174, 183
  conversion of, 186–188

  English, 166, 174
  English System of, 164
  gravitational, 174, 183–186
  hidden, 187
  International System of (SI), 165, 166, 174–183
  metric, 166
  supplementary, 175
urban planning, 15–16
U.S. Department of Commerce, 166

vacuum melting, 27
valves, expansion of, 30
vibrations, 90
viscosity, 169
volt, 180
volume, 164

watt, 180
weight, 164
work, 179
Wright, Wilbur and Orville, 7

zone refining, 26